Hocine MZAD

FUNDAMENTAL CONCEPTS OF CONDUCTIVE HEAT TRANSFER

Hocine MZAD

FUNDAMENTAL CONCEPTS OF CONDUCTIVE HEAT TRANSFER

Theoretical foundations and practical applications

ScienciaScripts

Cover image: www.ingimage.com

This book is a translation from the original published under ISBN 978-3-8416-3722-2.

Publisher:
Sciencia Scripts
is a trademark of
Dodo Books Indian Ocean Ltd. and OmniScriptum S.R.L publishing group

120 High Road, East Finchley, London, N2 9ED, United Kingdom
Str. Armeneasca 28/1, office 1, Chisinau MD-2012, Republic of Moldova, Europe
Managing Directors: Ieva Konstantinova, Victoria Ursu
info@omniscriptum.com

Printed at: see last page
ISBN: 978-620-8-87250-2

IN MEMORY OF MY FATHER,

Contents

TO MY WIFE,
AND MY CHILDREN.

Hocine MZAD, Professor,
Department of Mechanical Engineering,
Energy Speciality,
Faculty of Technology,
Badji Mokhtar University, Annaba.
O https://orcid.org/0000-0001-9900-6960
Please do not hesitate to send your comments on the book to the author's email address : hocine.mzad@univ-annaba.dz

PREAMBLE

This book does not claim to settle these debates. It seeks above all to offer a simple presentation of the main thermal theories to which they refer. However, it does not claim to be exhaustive; choices have had to be made, regrettably certainly, and priority has been given to the topicality of the issues. This is not to say that this approach is the alpha and omega of the discipline, but it remains the most sought-after in examinations and competitions, and this book is primarily aimed at students. It is therefore essential to provide clear, accessible explanations to make it easier to understand the key concepts. By focusing on the practical applications of thermal theories, this book aims to enhance students' preparation and help them excel in their studies.

Students of technology are often faced with a major difficulty: finding an equivalent in the subject studied to abstract concepts, presented in a very formalised way and which are not always intended to describe reality faithfully. However, it is entirely possible to understand certain physical mechanisms or phenomena using these theoretical tools. We have therefore tried here to make these sometimes obscure theories intelligible, and above all usable, while seeking not to sacrifice too much of the spirit of the models. As far as possible, we will also try to provide examples, data and reminders whenever possible.

This book was initially developed for the third year of the mechanical engineering degree course, but it is also intended for other science and technology specialities that study heat transfer. Engineers and masters students will also benefit from this book, which has a dual objective:

- To explore a number of concrete heat conduction problems in order to identify useful basic concepts.
- To assimilate the concept of heat conduction through a simple modelling approach.

Annaba, April 2025

H. Mzad

PART I

GENERAL CONCEPTS ABOUT THERMALS

Since ancient times, many people have been interested in steam and its use. They wanted to improve mechanical production, which at the time was limited to human or animal energy, but also to the natural force of elements such as wind and water.
Our industrial civilisation, which was born in the 17th century, was initially dominated by steam, so much so that today, a heating engineer is still often defined in industry as someone who works with boilers and steam systems. This century saw numerous attempts to improve the expansion force of steam.
At the time, the industrial revolution was seen as a race to achieve ever greater efficiency. James Watt's steam engine was one of the milestones of this revolution in terms of motive power. The steam engine was the first device to convert thermal energy into mechanical energy. Between 1776 and 1800, almost 500 machines were built, which goes to show the colossal impact Watt had on the Industrial Revolution.

1.1 The concept of heat

Alongside and in addition to fluid mechanics, thermal engineering is the discipline that enables engineers to understand and describe the operation of a wide range of industrial equipment, all of which involves the exchange of heat.
Thermodynamics is a specific scientific discipline within the broader field of physics. Thermodynamics is essentially based on principles that express the conservation of two fundamental quantities: energy and entropy. Energy exchanges fall into two categories: work, which results from the action of one or more forces applied to the system under consideration, and heat, which occurs naturally as soon as there is a temperature difference within the system under consideration or between it and its environment. From a more physical point of view, heat transfer originates in temperature differences. Energy is transferred in the form of heat whenever there is a temperature gradient within a system or when two systems at different temperatures are brought into contact.
The science of heat tells us about heat, analysing and quantifying its three modes of propagation: conduction, convection and radiation. Its

fundamental laws (Fourier, Newton and Stefan-Boltzmann) make it possible to predict the spatio-temporal evolution of temperature within physical systems and to calculate the heat exchanges between them.
In physics, heat is a special form of energy. The equivalence of heat and work is the first principle of thermodynamics. As a result, energy, work and heat all have the same unit: the joule.
Heat is transferred from one part of a substance to another, or from one body to another, in the form of the kinetic energy of disorderly molecular agitation. This transfer is the result of a difference in temperature between the two bodies. Heat spreads spontaneously from the body with the higher temperature to the one with the lower temperature, raising the temperature of the latter while lowering the temperature of the former, as long as the volume of the two bodies remains constant. This constitutes the second principle of thermodynamics, which highlights the notion of irreversibility: heat cannot propagate from a cold body to a hot body unless work is done.

1.2 The concept of temperature

The concepts of temperature and heat are not synonymous, as some common expressions might suggest. Historically, it was not until the 19th century that the concepts of temperature and heat were gradually explained in a satisfactory way at the microscopic level. In fact, the heat of a material body can be interpreted as being the sum of the energies of disordered agitation of the molecules (or atoms) that make it up. Temperature, on the other hand, measures the level of agitation: the greater the molecular agitation, the higher the temperature. The temperature of a system is an increasing function of the degree of thermal agitation of the particles, i.e. its thermal energy. It is defined by the heat transfer equilibrium with other systems.
When two bodies are placed in an adiabatic enclosure, the hotter body transfers heat to the colder body until the two bodies have the same temperature. Thermal equilibrium is then achieved.
Temperature is a thermodynamic property of a body and measures the microscopic agitation of matter. According to kinetic theory, the temperature of a body is a function of the average translational kinetic energy of its molecules (Fig. I.1). The kinetic energy of a body is zero at a temperature called absolute zero.

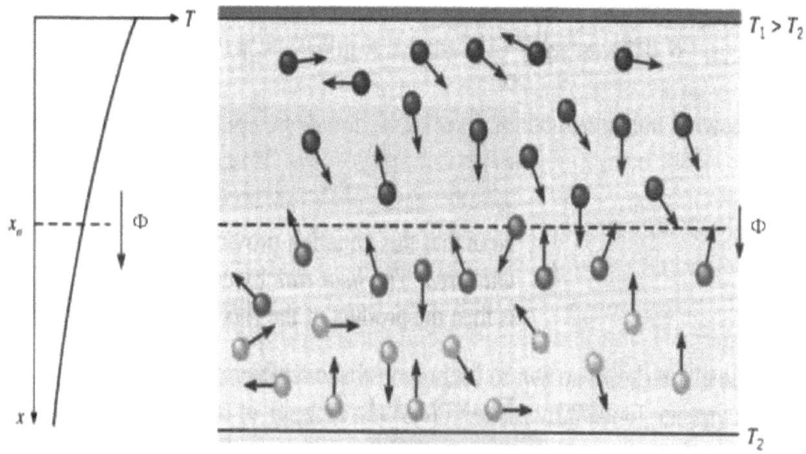

FIGURE I.1: Diagram of conduction in a fluid at rest.

1.3 Temperature measurements

The temperature of a body can be measured by observing the change in one of its specific properties, such as electrical resistivity or specific volume. For example, the mercury thermometer (Fig. I.2) measures the expansion of a column of mercury in a glass capillary, the variation in the length of the column being related to temperature.

The electrical resistance of a conductor or semi-conductor increases with temperature. This phenomenon is the basis of the resistance thermometer (Fig. I.3), in which a constant voltage, or electrical potential, is applied to the thermistor, the sensitive part of the sensor. For a thermistor of a given composition, the measurement of a given temperature induces a specific resistance for the thermistor. This resistance can be measured using a galvanometer and gives a temperature measurement.

Various thermistors made from nickel, manganese or cobalt oxides are used to measure temperatures between -46°C and 150°C. Similarly, thermistors using other metals or alloys can be used at higher temperatures: platinum, for example (up to 930°C). With suitable signal transformation circuits , the current measurement can be converted into a direct digital temperature measurement.

Highly accurate temperature measurements can be made with thermocouples, in which a small voltage difference (measured in millivolts) arises when two wires of different metals are joined to form a loop whose two connections are heated to different temperatures. Depending on the type of thermocouple, temperatures ranging from -

200°C to 2300°C can be measured. For very high temperatures and aggressive environments, the thermocouple must be protected in an alumina sheath (Fig. I.4). To increase the voltage signal, several thermocouples can be connected in series to form a thermopile. As the voltage depends on the difference between the temperatures of the links, one of these links must be maintained at a known reference temperature, otherwise an electronic compensation circuit must be integrated into the system to measure the actual temperature of the sensor.

Thermistors and thermocouples often have sensor sizes of less than 3 cm, enabling them to respond quickly to temperature variations and making them ideal for many applications in biology and engineering.

The optical pyrometer is used to measure the temperatures of solid objects above 700°C (Fig. I.5). At such temperatures, solid objects emit sufficient energy in the visual range to allow optical measurement by exploiting the apparent temperature phenomenon of incandescent colour, the colour emitted by incandescent objects ranging from dark red to yellow to white at around 1300°C. The pyrometer contains a filament similar to that of an electric bulb, controlled by a calibrated rheostat, so that the colours of the incandescent filament correspond to precise temperatures. Thus, the temperature of a glowing object can be measured by looking at the object through the pyrometer and adjusting the rheostat until the filament becomes invisible against the background of the object: then the temperatures of the filament and the object are equal and can be read on the rheostat scale.

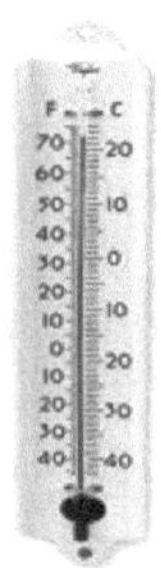

FIGURE I.2: Mercury thermometer. **FIGURE I.3:** Resistance thermometer.

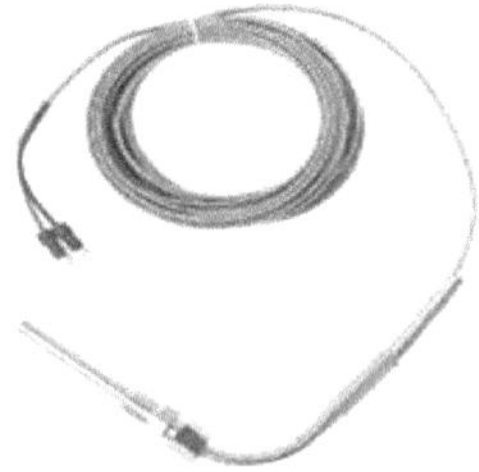

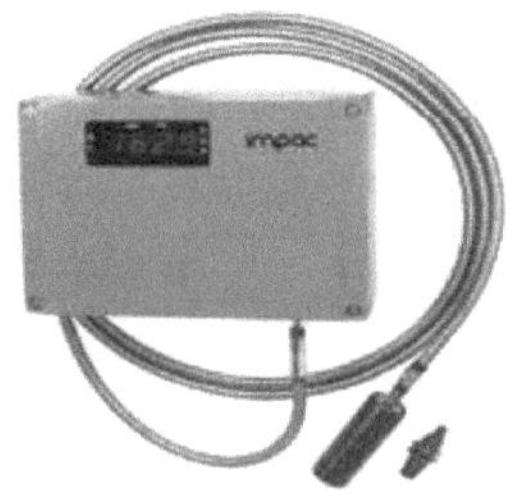

FIGURE I.4: Type K thermocouple. **FIGURE I.5**: Infrared optical pyrometer.

1.4 Thermometric scales

Thermometric scales are the scales used to determine the temperature of a body. A temperature scale is a method of expressing temperature in numerical form. Various scales have been used to measure temperature since 1700. Among these, we will list some of the most common temperature units (Fig. I.6):

■ The Kelvin, often used in science and technology, is the basic unit of temperature in the SI system (International System of Units). The abbreviation for the unit kelvin is K (no degree or degree sign). The kelvin unit was first introduced in 1848 by William Thomson (Lord Kelvin). The Kelvin unit was initially developed by moving the starting point of the Celsius unit to absolute zero. With the redefinition of the International System of Units (SI) in 2019, the Kelvin unit was defined by setting the numerical value of Boltzmann's constant at 1.380649×10^{-23} J/K.

■ Celsius is currently a derived unit of temperature in the SI system, Kelvin being the base unit. The abbreviation for Celsius is °C (degree Celsius).

The size of a degree Celsius is the same as that of a Kelvin, and 0 K is equal to -273.15 °C. It was the Swede Andreas Celsius who first introduced the Celsius unit and scale in 1742. The two main reference points for the Celsius scale were the freezing point of water (or the melting point of ice) defined at 0°C and the boiling point of water at 100°C. Celsius is better suited to everyday use than kelvin and is very popular around the world, although not widely used in the US. One degree Celsius is sometimes called centigrade.

■ The Fahrenheit scale was first introduced by a Dutchman called Gabriel Fahrenheit in 1724. The abbreviation for the Fahrenheit unit is °F. The main reference points of the scale are the freezing point of water defined at 32 °F, the boiling point of water exactly 212 °F and the temperature of the human body is about 98 °F. In many areas, Fahrenheit has been replaced by Celsius as the unit of temperature.

■ The Rankine scale was first introduced in 1859 by the Scotsman William Rankine. The abbreviation for Rankine is °R or °Ra. The reference point of the Rankine scale is absolute zero at 0 °Ra, as on the Kelvin scale. The size of a degree Rankine is the same as the size of a degree Fahrenheit, but the zero point is very different. The freezing point of water is 491.67 °Ra. The Rankine scale is not widely used. It was used in certain fields of technology in the United States.

In the SI system, the basic thermal unit is the Kelvin, a thermodynamic temperature unit. Until 1954, the temperature scale was defined as the difference of 100 degrees between the freezing and boiling points of water. This imprecise definition was replaced in 1967 by the current definition, which uses a fundamental fixed point: the triple point of water. Thus, the kelvin is equal to 1/273.16 of the thermodynamic temperature of the triple point of water. This point describes the equilibrium pressure and temperature conditions of water in its three states: liquid, solid and gas.

The absolute temperature coincides with the thermodynamic scale temperature. In cryogenics, where temperatures as low as 0.00001 K are sought, paramagnetic materials are demagnetised. At the other end of the scale, momentary temperatures in excess of 100 million K have been observed during nuclear explosions.

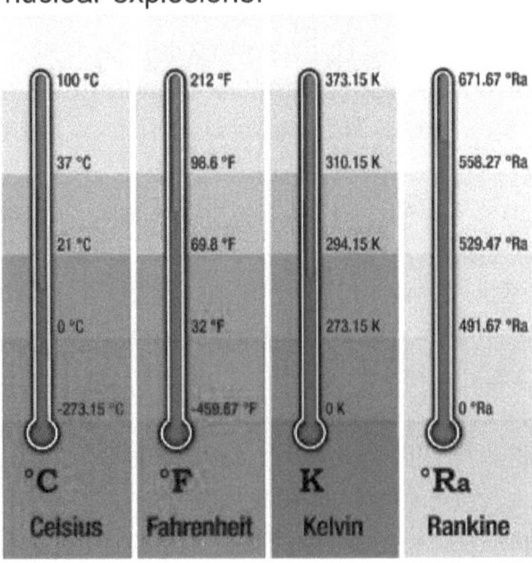

FIGURE I.6: Conversions between common temperature units.

1.5 Heat units

We have seen that in physics, the quantity of heat is expressed in the

same units as energy and work, i.e. in joules (J). We also use the calorie (Cal), defined as the amount of heat needed to raise the temperature of 1 g of water from 14.5 °C to 15.5 °C under a pressure of 1 atm.
Mechanical energy can be converted into heat by friction, and the mechanical work required to produce one calorie is called the mechanical equivalent of the calorie, and we have :

4.1855 J = 1 Cal

From the calorie, other units of heat quantity are sometimes introduced:

1 kilocalorie (kcal) = 1000 cal

1 thermie = 1000 kcal

1 kilowatt-hour (kWh) = 3.6×10^6 J = 860112 cal

1 kWh = 0.86 thermie

According to the law of conservation of energy, all the mechanical work done to produce heat by friction is converted into energy in the bodies involved in that work. This fact was first conclusively demonstrated by Joule, who heated water in a closed container with paddle wheels and thus proved that the rise in temperature of the water was proportional to the work exerted by turning the wheels. In an internal combustion engine, where heat is transformed into mechanical energy, the law of conservation of energy also applies. However, some of the energy is always lost or dissipated in the form of heat, because no engine is 100% efficient.

1.6 The concept of heat capacity

The heat capacity (also called mass heat or mass heat capacity) of a substance, designated by the letter *C*, is a characteristic property. It gives the precise capacity of this substance to absorb or release heat. It is determined by the amount of energy required to raise the temperature of a unit mass of a substance by one degree. In other words, the mass heat capacity is the amount of heat that one gram of a substance must absorb to raise its temperature by one degree Celsius. It can also be the amount of heat that one gram of the substance must lose for its temperature to drop by one degree Celsius.
According to the law formulated by French chemists Pierre Louis Dulong and Alexis Thérèse Petit, the heat capacities of solid elements are inversely proportional to their atomic masses. This means that the product of the mass heat capacity and the atomic mass is constant for all solid elements.
As a heated gas expands, additional heat is required to raise its temperature, since some of this energy is used to expand the gas. This

is why, in the case of a gas, a distinction must be made between the mass heat capacity at constant pressure *Cp* and the mass heat capacity at constant volume *Cv*. The mass heat capacity at constant pressure will always be greater than the mass heat capacity at constant volume.
In the International System (SI) of units, the mass heat capacity is expressed in J/(kg.K). If we consider a solid body *S* of mass *m* and mass heat capacity *Cp*, and if we supply it with a quantity of heat ΔQ, the increase in the heat contained by the body is responsible for a rise in temperature Δ *T*, given by the fundamental relationship :

$$\Delta Q = m\, C_p\, \Delta T \qquad (I.1)$$

With :
ΔQ in Joules,
m in kilograms,
Cp in J/(kg.K) or J/(kg.°C),
Δ *T* in °C or K.
Appendix 1 contains lists of the heat capacities of various elements and liquid substances, measured under standard conditions at room temperature and normal pressure (TAPN), i.e. at 25°C and 101.3 kPa.

1.7 Notion of enthalpy

In thermodynamics, enthalpy is the state function of a system defined by the relation :

$$H = U + pV \qquad (I.2)$$

This function is expressed in Joules per kg (J/kg).
U is the internal energy of the system per unit mass.
p and *V* are the pressure and volume of the system, respectively.
Enthalpy is independent of the way in which a system reaches its state. It is based on the initial situation and the final situation without considering intermediate states. During a transformation at constant pressure, the heat received by the system corresponds to the variation in its enthalpy.
Consider a system of volume *V* in an initial state characterised by its pressure *p*, temperature *T* and enthalpy *Hi*. By supplying it with a quantity of heat ΔQ, let us pass this system into a final state which will be characterised by the same pressure *p*, the same volume *V*, its final temperature *Tf and* its enthalpy *Hf*. The enthalpy of this system has undergone a change :

$$\Delta H = H_i - H_f = \frac{\Delta Q}{m} = C_p\, \Delta T \qquad (I.3)$$

For a thermic engineer, enthalpy is nothing other than the heat contained

per unit mass in a solid or per Nm^3in a gas.
Enthalpy is particularly important when describing transformations at constant pressure. This is the case for most chemical reactions that take place at atmospheric pressure. The enthalpy variation is then equal to the quantity of heat released or absorbed by the system, depending on whether the reaction is exothermic or endothermic. An exothermic reaction is characterised by the fact that it produces enthalpy and an endothermic reaction by the fact that it consumes enthalpy.

1.8 Latent heat of change of state

In thermodynamics, the latent heat of change of state (also known as the enthalpy of change of state) of a pure substance is, by definition, the variation in enthalpy that accompanies a change of state of the substance in relation to the quantity of matter involved in this transformation. Latent heat changes the physical state of a substance. As opposed to sensible heat, which changes the temperature of a substance.
The phase of a pure substance is the solid, liquid or gaseous state it assumes according to conditions of temperature and pressure. The passage of a substance from a solid state to a gaseous state is called sublimation, from a solid state to a liquid state, fusion, and from a liquid state to a vapour state, evaporation or vaporisation. If the pressure is constant, these processes take place at a constant temperature. A certain amount of heat is required to achieve these three changes of state. This is called latent heat. Thus, there are latent heats of sublimation, fusion and vaporisation.
When water is boiled in an open container at atmospheric pressure, the temperature does not rise above 100°C, even if the amount of heat is increased. The heat absorbed by the water without any change in temperature is called latent heat and is used to transform the water into steam. It is recovered when the steam is condensed to form water.
In a similar way, if a mixture of water and ice is heated in a glass, its temperature does not change until all the ice has melted. The heat absorbed is used to melt the ice.
For example, the heat of vaporisation of a litre of water is 2257 kJ/kg (at atmospheric pressure and 100°C). That's 5.4 times more than it takes to heat a litre of water from 0 to 100°C! Note that the heat of vaporisation varies according to the temperature of the water being evaporated: from 2257 kJ/kg at 100°C, the heat of vaporisation is 2454 kJ/kg at 20°C and 2501 kJ/kg at 0°C. It is therefore slightly easier for a drop to go into the

vapour state when it is already at 100°C.
Certain materials are selected for their high latent heat at a given temperature level: these are known as phase-change materials, or phase-change salts.

1.9 Heat flux

The amount of heat ΔQ transferred per unit time is called heat flow Φ :

$$\Phi = \frac{\Delta Q}{\Delta t} \qquad (I.4)$$

Φ is a heat flow rate, and is expressed in joules per second, i.e. watts. Heat transfers are measured by flows and have the same dimensions as power, for example electrical power. The following table (Tab. I.1) summarises the different mechanical and thermal units encountered so far.

Physical quantity	Unit	Notation
Energy Work Quantity of heat	Joule	J
Quantity of heat	Calorie, thermie	Cal, th
Energy	Watt-hour	Wh
Power Thermal flux	Watt	W
Temperature	Degree Celsius, Kelvin	°C, K
Heat capacity Enthalpy Entropy	Joule/Kelvin	J/K
Heat capacity by mass	Joule/(kilogram.Kelvin)	J/(kg.K)

TABLE I.1: Thermal units.

1.10 The concept of thermal conductivity

The thermal conductivity of a material reflects its ability to transmit heat by conduction. Its value is given by the coefficient (λ). It can be expressed in W/(m.°C) or W/(m.K).
The higher the thermal conductivity, the more conductive the material. Conversely, the lower the coefficient of thermal conductivity, the more insulating the material. Metals such as gold, silver and copper have a high thermal conductivity, while materials such as glass and asbestos have a much lower thermal conductivity.
The technical and commercial documentation for the materials must specify the conditions under which this value is obtained, together with the value of λ. We will try to approach the normal values for use (temperature of 10°C to 20°C). In general, the thermal values of materials are given as a function of their dry density. For certain families of materials, several useful thermal conductivity values are given as a function of the density of the material. If this is not known, the highest of

the useful thermal conductivity values given for the family in question is adopted.
The thermal conductivity coefficient of a material varies according to its temperature and humidity. This coefficient is only valid for homogeneous materials. It has no meaning for heterogeneous materials, through which heat propagates at the same time by conduction, convection and radiation.
In many industrial applications, it is necessary to know the speed of heat propagation in a solid for a given temperature difference. Sophisticated mathematical calculation techniques are used, using computers.
It can be said that there is no area of engineering today that is not confronted with heat transfer problems, which makes knowledge of thermics increasingly necessary, if not strategic. Examples include the space industry, aeronautics and automotive engines,
lighting, on-board electronics, petrochemicals, food processing, waste treatment, construction, comfort and the production of electrical energy from hydrocarbon combustion or nuclear reactions. The control and optimisation of systems in which thermal energy plays an obvious role (furnaces, heat exchangers turbines, computers, transformers
transformers, radiators, condensers, heaters, evaporators, braking systems, etc.) are currently the subject of a great deal of research and development work.
Finally, only thermal conduction, the only possible mode of heat transfer in solids, will be discussed in this manuscript. The exact mechanism of conduction in solids is still not fully understood, but is mainly due to the movement of free electrons in the body, which is triggered as soon as a temperature difference is established. So good conductors of heat are generally good conductors of electricity.

PART II

CONDUCTION HEAT

11.1 Introduction

Heat transfer is the process by which energy is exchanged in the form of heat between bodies or media at different temperatures T_1 and T_2. From a more physical point of view, heat transfer originates in temperature differences. Energy is transferred in the form of heat whenever there is a temperature gradient within a system, or when two systems at different temperatures are brought into contact by a single relationship that results from a combination of independent physical mechanisms.

Conduction is defined as the mode of heat transfer caused by the difference in temperature between two regions of a solid, liquid or gaseous medium. The observable macroscopic effect is an equalisation of the system's temperatures. However, if certain zones are kept at a constant temperature by heat input (heat reservoir) or heat output (heat sink), there is a continuous transfer of heat from the hot region to the cold region.

In fact, heat can be transmitted by conduction, convection or radiation. Although all three processes can occur simultaneously, one of the mechanisms is generally dominant. For example, heat is transmitted mainly by conduction through the brick walls of a house; water in a saucepan on a cooker is heated mainly by convection; the earth receives its heat from the sun largely by radiation.

Conduction mainly concerns solids. Although it also occurs in liquids and gases, it is usually negligible compared with convection and radiation.

The fundamental problem of conduction is to find the temperature at any point in the thickness of the floor. To solve these problems, it is necessary to know the local mechanism of the phenomenon. We will see that this mechanism will be described by an equation, known as the heat equation, which links together the various variables involved (temperature T, time t, space variables x, y, z).

11.2 Thermal quantities

11.2.1 Thermal field

The thermal field of a body is the set of temperatures at any point on the body at a given time: $T(M, t)$, where $M(x, y, z)$ refers to any point on the body. This temperature field can be time-independent (permanent) or time-dependent (variable). The locus of points M having the same temperature at a time t is called the isothermal surface.

Assume a source of internal energy which supplies the body with a quantity of heat *dQ* for time *dt*. The heat flow, the equivalent of power in watts, which will propagate towards the environment is :

$$\Phi = \frac{dQ}{dt} \qquad \text{(II. 1)}$$

Assuming that the structure of the body is perfectly uniform, we can simply reason about a unit area, and then the results obtained can be extended to the entire surface of the body *S*.

The power exchanged per unit area is called the heat flux density, i.e. the quantity :

$$\varphi = \frac{dQ}{S\,dt} = \frac{\Phi}{S} \qquad \text{(II. 2)}$$

The above assumption of uniformity of the body is simplistic insofar as we consider only one heat flux density φ valid at any point on the surface *S* of the body. We need to move on to a local description. To do this, let us introduce a vector field $\varphi(M, t)$ representing the local heat flux density at each point $M(x, y, z)$ of the material system.

This vector $\varphi(M, t)$ characterises, at each point of the medium, the direction, sense and intensity of the heat flux caused by the presence of the internal source. The modulus of such a vector has the dimension of a power per unit area, and is therefore expressed in W/m^2.

If we consider an elementary surface *dS* surrounding the point *M* oriented by its unit normal $\vec{n}$ (Fig. II.1), the direction of the local heat flux density will make an angle a with the normal *n*. The elementary heat flux $d\Phi$ through the surface *dS* will be given by the expression :

$$d\Phi = |\vec{\varphi}(M,t)|\,\vec{n}\;dS = |\vec{\varphi}(M,t)|\cos\alpha\;dS \qquad \text{(II. 3)}$$

this quantity represents energy per unit time and is therefore expressed in watts.

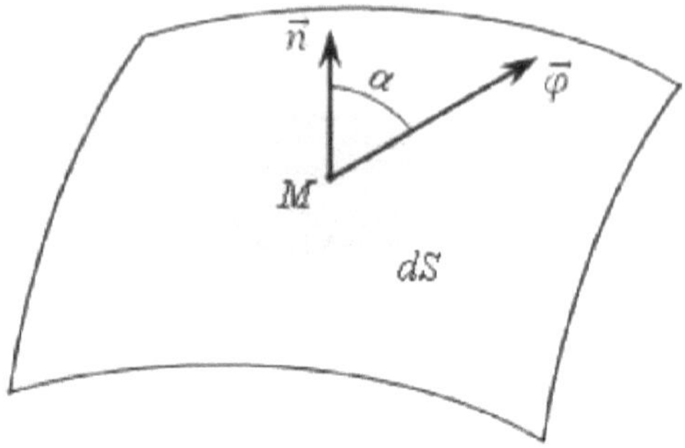

FIGURE II.1: Local heat flux density.

A parallel can be drawn between the flux density vector field in thermics and the velocity field in fluid mechanics. For each of these two vector fields, current lines can be defined which are the envelopes of the vectors considered. In each of the two disciplines, fluid mechanics and heat engineering, the set of streamlines resting on a closed contour constitutes a stream tube.

11.2.2 Fourier's law

In 1822, the French mathematician Joseph Fourier gave a precise mathematical definition of conduction. According to Fourier's law, the speed at which heat is conducted through a body per unit cross-section is proportional to the opposite of the body's temperature gradient. The proportionality factor is the thermal conductivity of the material.

Fourier likened heat conduction to the flow of a fluid from hot to cold regions, the only manifestations of which in matter are temperature variations (macroscopic effect). The dilatations of the devices will be neglected.

Statement: There is a linear relationship between the effect of interest, i.e. the heat flux density $\vec{\varphi}(M,t)$ and its cause, which is the existence of a temperature gradient $\overrightarrow{grad}\,T$. In other words, at any point in an isotropic medium the instantaneous heat flux density is proportional to the temperature gradient, which translates into :

$$\vec{\varphi}(M,t) = -\lambda\,\overrightarrow{grad}\,T \qquad \text{(II.4)}$$

The proportionality coefficient λ is called the thermal conductivity of the medium and is expressed in W/(m.K). For an isotropic and homogeneous medium, λ depends only on the temperature *T*. In many

practical cases, where temperature differences are not too great, λ can be considered a constant for a given medium with sufficient accuracy.
By convention, $\vec{\varphi}(M,t)$ is counted positively in the direction of heat flow, i.e. towards decreasing temperatures. $\overrightarrow{grad}\,T$. is a vector carried by the same axis, but in the opposite direction to $\vec{\varphi}(M,t)$, hence the minus sign of Fourier's law.

Material		λ (Values at room temperature)
Silver		417
Copper		380
Aluminium		230
Metals and alloys	Brass	110-90
Pure iron		72
Steel		52
Lead		35
Stainless steel		15
Building materials	Solid concrete	1,75
	Vermiculite concrete	0,31
	Cellular concrete	0,24
	Refractory bricks	0,11 -0,61
	Gypsum	0,43
	Wood	0,15
	Cork compensated	0,1
	Glass	1,15
Liquids	Water	0,56
	Alcohol	0,13
Gas	Air	0,02
	Water vapour	0,016
	Carbon dioxide	0,014
Insulation	Expanded cork	0,043
	Glass wool	0,035
	Wool	0,035
	Asbestos	0,15
	Asbestos cement	0,8
	Polyurethane	0,03
	Polystyrene	0,04
	Polyvinyl chloride	0,031

TABLE II.1: Thermal conductivity of some materials.

In addition to temperature, thermal conductivity also depends on the state of the material under consideration (solid, liquid or gas). Table II. 1 gives some values of λ for various materials at room temperature.

II.2.3 Orthogonality of gradient and isotherm

At any point *M* in the medium, at any instant :

$$dT = \overrightarrow{grad}\, T\, \overrightarrow{dM} \qquad (II.5)$$

If we consider an elementary displacement *dM* on the isotherm passing through *M*, along this isotherm we have :

$$dT = 0 = \overrightarrow{grad}\, T\, \overrightarrow{dM} \qquad (II.6)$$

the vectors $\overrightarrow{grad}\, T$ and $\overrightarrow{dM}$ are orthogonal.

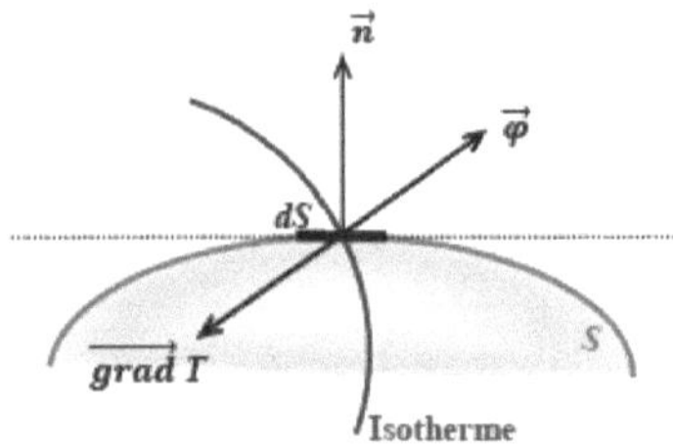

FIGURE II.2: Orthogonality of the gradient and ¡'isotherm.

The temperature gradient at each point is normal to the isothermal surface passing through that point. It follows that the streamlines, to which the heat flux density vectors are tangent, are also normal to the isothermal surfaces (Fig. II.2). The walls of a current tube are therefore normal to the isotherms. No flow therefore passes through them. These walls are said to be adiabatic.

So, in steady state, the heat flow is conservative in a flow tube. This property is the exact analogy of the constancy of flow in a flow tube in fluid mechanics.

II.2.4 Conductance and thermal resistance of a wall

Consider a homogeneous wall of thickness *e*, constant surface area *S*, and thermal conductivity λ (Fig. II.3). The face at $x = 0$ is maintained at temperature T_1 and the face at $x = e$ at temperature T_2. In steady state, the heat flow Φ through this wall is such that :

$$\Phi = -\lambda\, S\, \frac{dT}{dx} \qquad (II.7)$$

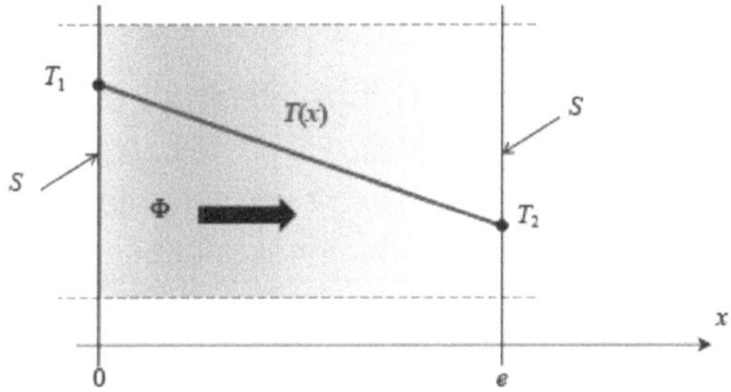

FIGURE II.3: Thermal resistance of a plane wall.

In this case, the temperature distribution is linear:

$$T(x) = -\frac{T_1 - T_2}{e}\, x + T_1 \qquad (II.8)$$

The temperature gradient is then constant, i.e. :

$$\frac{dT(x)}{dx} = -\frac{T_1 - T_2}{e} \qquad (II.9)$$

The relation (II.7) then becomes :

$$\Phi = \frac{\lambda\, S}{e}\,(T_1 - T_2) \qquad (II.10)$$

The heat flow is therefore proportional to the temperature difference between the faces of the wall. The proportionality coefficient represents the thermal conductance of the medium:

$$C_t = \frac{\lambda\, S}{e} \qquad (II.11)$$

We also define the thermal resistance of the wall :

$$R = \frac{e}{\lambda\, S} \qquad (II.12)$$

Under these conditions, the temperature difference between the two sides of the wall is written as :

$$\Delta T = T_1 - T_2 = R\,\Phi \qquad (II.13)$$

We can see that there is a proportional relationship between the heat flux transmitted by conduction through the wall and the temperature difference T_1 - *T2*. This relationship naturally leads to an electrical analogy. Ohm's law for a homogeneous cylindrical conductor through which an electric current of intensity *I* flows is written :

$$\Delta V = V_1 - V_2 = R\, I \qquad (II.14)$$

in which the electrical resistance :

$$R = \frac{\rho}{L\,A}$$

is the resistivity of the conductor (e.g. an electric wire), *L* is its length, and *A* is its cross-sectional area.

The form of the electrical resistance *R* is reminiscent of the expression giving the thermal resistance of the wall (Eq. II.12). The inverse of the conductivity is the resistivity, $p = 1/\lambda$. The result is the table of analogies below (Tab. II.2) between thermal and electrical quantities.

Thermal quantity	**Electrical quantity**
Temperature difference ΔT (K)	Potential difference ΔV (V)
Heat flux Φ (W)	Current intensity *I* (A)
Thermal resistance *R* (K/W)	Electrical resistance *R* (Ω)
Electric field (V/m)	Temperature gradient (K/m)

TABLE II.2: Thermal/electrical analogies.

We will see throughout this chapter that the application of the laws of electricity to the series or parallel combination of these thermal resistances provides a simple and convenient method of solving a large number of steady-state thermal problems.

II.3 Steady state conduction

For the sake of clarity, we will start by thinking in terms of a one-dimensional problem. Such a one-dimensional problem is known as the wall problem, i.e. a medium bounded by two parallel planes in which heat propagates only along the normal to these planes. The temperature gradient is therefore carried by this normal. The isotherms are planes parallel to the faces. Such a thermal field is unidirectional. In this case, the temperature *T* is a function only of the abscissa *x*.

Let's consider a wall of thickness *e* separating two domains with respective temperatures T_1 and T_2, with T_1 greater than T_2 (Fig. II.4).

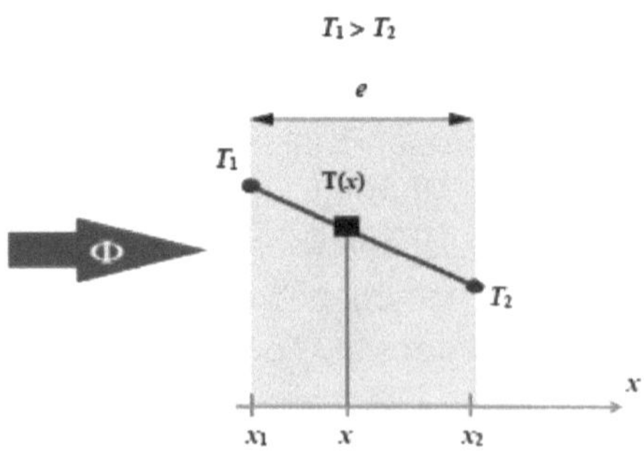

FIGURE II.4: Temperature linearity.

If the phenomenon under consideration is in steady state, the temperatures no longer change. This means that a state of equilibrium has been reached between the heat input through the hot side of the wall and the heat output through the cold side, which translates into :

Incoming flow = Outgoing flow

$$S\,\varphi_1 = S\,\varphi_2$$

Assuming an imbalance, if :

$S\,\varphi_1 - S\,\varphi_2 = S\,\Delta\varphi > 0$

The amount of excess heat $S\ \Delta\varphi$ accumulated each second in the wall, is related to the physical characteristics of that wall by the expression:

$$S\,\Delta\varphi = \frac{m\,C_p\,\Delta T}{\Delta t} \qquad \text{(II.15)}$$

And it is this accumulation of heat that leads to an increase in the temperature of the wall. In steady state, however, we have :

$$m\,C_p\,\frac{\Delta T}{\Delta t} = 0 \qquad \text{(II.16)}$$

This is immediately integrated to give the temperature profile in the wall, which varies linearly as a function of *x* :

$$m\,C_p\,T(x) = ax + b \qquad \text{(II.17)}$$

To determine the values of the constants *a* and *b* which will define a particular solution, several types of surface conditions can be applied to this wall, depending on the nature of the physical problem to be solved.

II.3.1 Simple wall with isothermal faces

One possible configuration is to imagine that each wall is maintained at a constant, known temperature. This would be the case, for example, for a house that we want to heat to an indoor temperature T_1 while the outdoor temperature is T_2. From the boundary conditions (Fig. 11.4) and equation (II.17), we have :

For $x = 0$, $T(x=0) = T_1$: $\quad m\,C_p\,T_1 = b$

For $x = e$, $T(x=e) = T_2$: $\quad m\,C_p\,T_2 = ae + b$

From these two equations we can easily derive :

$$a = \frac{m\,C_p\,(T_2 - T_1)}{e}$$

By replacing the constants *a* and *b* in Eq. (II.17), the temperature profile in the wall will have the expression :

$$m\, C_p\, T(x) = \frac{m\, C_p\, (T_2 - T_1)}{e} x + m\, C_p\, T_1 \qquad \text{(II.18)}$$

Let:

$$T(x) = T_1 - \frac{(T_1 - T_2)}{e} x \qquad \text{(II.19)}$$

The previous equation confirms that the temperature profile is linear, as shown in figures (II.3) (II.4), with a gradient equal to $-\frac{(T_1-T_2)}{e}$.
The heat flow density through the wall is given by Fourier's law :

$$\varphi = -\lambda \frac{dT}{dx} \qquad \text{(II.20)}$$

As the gradient is constant, this heat flow has a constant value whatever the abscissa *x* in the wall. The temperature will therefore not vary as a function of time. This is the steady state. Equation (II.20) can then be written in terms of finite differences :

$$\varphi = -\lambda \frac{(T_1 - T_2)}{(x_1 - x_2)} \qquad \text{(II.21)}$$

i.e. :

$$\varphi = \lambda \frac{(T_1 - T_2)}{e} \qquad \text{(II.22)}$$

The heat flux transmitted by conduction through the wall is therefore :

$$\Phi = S\, \varphi = \lambda\, S \frac{(T_1 - T_2)}{e} \qquad \text{(II.23)}$$

II.3.2 Wall in contact with two fluids

Now consider a wall of thickness *e* in contact with two fluids of constant temperatures *T* "1 and *T* "2. Between the wall and the two fluids on either side, heat is exchanged by convection. The respective temperatures of each of the two sides of the wall are T1 and T2. This is what is known as a mixed problem, combining two modes of heat transfer: convection and conduction.

Since the problem is one-dimensional, the conservation of heat flow (stationary problem) results in the equality of the following flows:

Φ1, flow given up by the hot fluid to the wall by convection.

Φ, flow through the wall by conduction, Eq. (II.23).

Φ2, flux given up by the wall to the cold fluid by convection. i.e. :

$$\Phi_1 = \Phi = \Phi_2 \qquad (II.24)$$

According to Newton, the convective fluxes Φ1 and Φ2 are defined such that :

$$\Phi_1 = h_1\, S\, (T_{\infty 1} - T_1) \qquad (II.25)$$

$$\Phi_2 = h_2\, S\, (T_2 - T_{\infty 2}) \qquad (II.26)$$

h 1 and *h* 2 are convective heat exchange coefficients expressed in W/(m(2)·K).

In addition to the conductive thermal resistance defined above, there are the convective thermal resistances: expressed in K/W as the conductive thermal resistance.

$$R_{c1} = \frac{1}{h_1\, S} \qquad (II.27)$$

$$R_{c2} = \frac{1}{h_2\, S} \qquad (II.28)$$

Given equations (II.23), (II.25) and (II.26) :

$$\begin{cases} (T_{\infty 1} - T_1) = \dfrac{\Phi}{h_1\, S} \\ (T_1 - T_2) = \dfrac{\Phi\, e}{\lambda\, S} \\ (T_2 - T_{\infty 2}) = \dfrac{\Phi}{h_2\, S} \end{cases} \qquad (II.29)$$

The combination of these three equations gives :

$$(T_{\infty 1} - T_{\infty 2}) = \Phi \left(\frac{1}{h_1\, S} + \frac{e}{\lambda\, S} + \frac{1}{h_2\, S}\right) \qquad (II.30)$$

i.e. :

$$(T_{\infty 1} - T_{\infty 2}) = \Phi\, R_{eq} \qquad (II.31)$$

where *Req* is the equivalent (total) resistance of three resistors in series, such that :

$$R_{eq} = R_{c1} + R + R_{c2} \qquad \text{(II.32)}$$

It should be noted, however, that the surface area *S* is constant in this case (plane wall with identical internal and external surfaces). We will see that this is not always the case, particularly for cylindrical and spherical shapes.

According to Newton's law, equation (II.31) can be written as follows:

$$\Phi = U\,S\,(T_{\infty 1} - T_{\infty 2}) \qquad \text{(II.33)}$$

U designates an overall exchange coefficient which has the expression :

$$U = \frac{1}{(R_{c1} + R + R_{c2})\,S} \qquad \text{(II.34)}$$

Consider a wall of thickness *e,* thermal conductivity λ, and whose end faces (at the boundaries) are in contact with two fluids at temperatures *T* "1 and *T* "2 (Fig. II.5). Under steady-state conditions, the heat flow is conservative:

$$\Phi_x = \Phi_{x+dx} = \Phi \qquad \text{(II.35)}$$

or :

$$-\lambda\,S\left(\frac{dT}{dx}\right)_x = -\lambda\,S\left(\frac{dT}{dx}\right)_{x+dx} = -\lambda\,S\left(\frac{dT}{dx}\right) \qquad \text{(II.36)}$$

If the flow is unidirectional, the temperature at a point *x* is given by :

$$T_{\infty 1} - T(x) = \big(R_{c1} + R(x)\big)\,\Phi \qquad \text{(II.37)}$$

where :

$$-\lambda\,S\left(\frac{dT}{dx}\right)_x = -\lambda\,S\left(\frac{dT}{dx}\right)_{x+dx} = -\lambda\,S\left(\frac{dT}{dx}\right) \qquad \text{(II.36)}$$

By replacing Φ by its expression from equation (II.31), equation (II.37) becomes :

$$T(x) = T_{\infty 1} - \big(R_{c1} + R(x)\big)\frac{T_{\infty 1} - T_{\infty 2}}{R_{c1} + R + R_{c2}} \qquad \text{(II.39)}$$

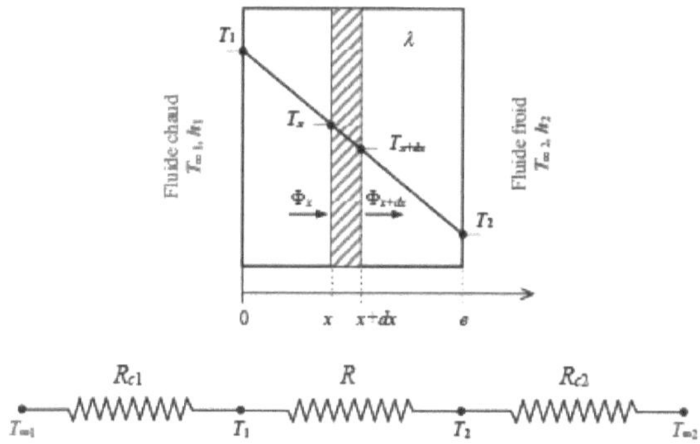

FIGURE II.5: Heat conduction in the presence of two fluids.

In general, for a wall made up of *n* layers in perfect contact, such as a multilayer wall (Fig. II.6), the heat flow (thermal losses) can be deduced:

$$\Phi = \frac{T_{\infty 1} - T_{\infty 2}}{R_{c1} + \sum_{i=1}^{i=n} R_i + R_{c2}} \qquad (II.40)$$

with :

$$R_i = \frac{e_i}{\lambda_i\, S}$$

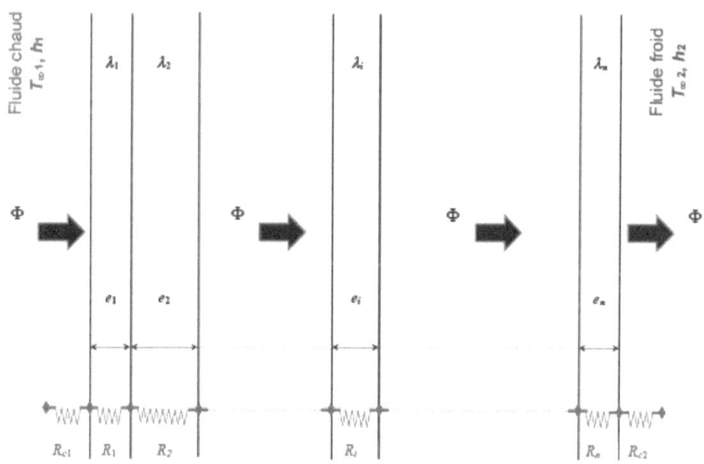

FIGURE II.6: Multi-layer wall.

II.3.3 Composite wall

As an example, consider an external insulation or soundproofing wall made of hollow blocks (Fig. II.7). This is the case most often encountered in reality, such as the double-walled walls of the facades of houses, made up of bricks separated by a layer of air.

Assuming unidirectional transfer and taking into account the axes of symmetry, we can return to the calculation of the flow through the isolated element on the right of Figure II.7. We propose to calculate the equivalent thermal resistance *Req* of a portion of wall of width *w* and height $\ell = \ell_1 + \ell_2 + \ell_3$, using the laws of association of resistances in series and in parallel by the equation :

$$R_{eq} = R_{c1} + R_1 + \frac{1}{\frac{1}{R_3} + \frac{1}{R_4} + \frac{1}{R_5}} + R_2 + R_{c2} \qquad \text{(II. 41)}$$

equation in which each resistance has the expression :

$$R_{c1} = \frac{1}{h_1(w \times l)}\ ;\ R_1 = \frac{e_1}{\lambda_1(w \times l)} = R_2\ ;\ R_3 = \frac{e_2}{\lambda_2(w \times l_1)}\ ;\ R_4 = \frac{e_2}{\lambda_1(w \times l_2)}\ ;$$

$$R_5 = \frac{e_2}{\lambda_2(w \times l_3)}\ ;\ R_{c2} = \frac{1}{h_2(w \times l)}$$

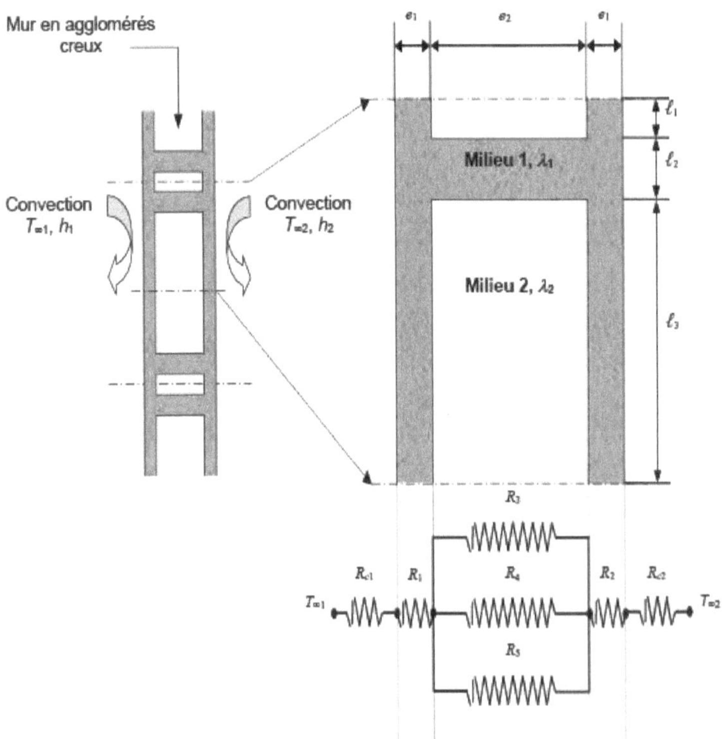

FIGURE II.7: Composite wall.

II.3.4 Hollow cylinder (tube)

Consider a hollow cylinder of length *L*, thermal conductivity λ, inner

radius *r* and outer radius *re,* with the walls of the cylindrical surfaces at uniform and constant temperatures T_1 and T_2 (Fig. II.8).
The cylinder is assumed to be sufficiently long in relation to the radii so that the longitudinal temperature gradient is negligible compared with the radial gradient. The isotherms are coaxial cylindrical surfaces.
Since the problem is radial, the temperature is only a function of the radius *r*. Remember that the Laplacian in cylindrical coordinates (independent of 0and *z*) is written as :

$$\Delta T = \frac{d^2T}{dr^2} + \frac{1}{r}\frac{dT}{dr} \qquad (II.42)$$

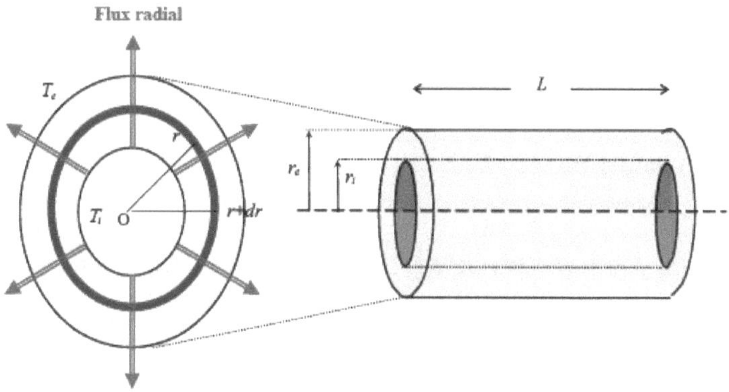

FIGURE II.8: Conduction in a hollow cylinder.

or :

$$r\,\frac{d^2T}{dr^2} + \frac{dT}{dr} = 0 \qquad (II.43)$$

or :

$$\frac{d}{dr}\left(r\,\frac{dT}{dr}\right) = 0 \qquad (II.44)$$

after a first integration :

$$r\,\frac{dT}{dr} = \mathrm{A} \Rightarrow \frac{dT}{dr} = \frac{\mathrm{A}}{r}$$

hence :

$$T = \mathrm{A}\ln r + \mathrm{B} \qquad (II.45)$$

To find the constants A and B, consider the Dirichlet boundary conditions :

$$\begin{cases} T = T_e \text{ si } r = r_e \\ T = T_i \text{ si } r = r_i \end{cases}$$

We obtain the system :

$$\begin{cases} T_e = A\ln r_e + B \\ T_i = A\ln r_i + B \end{cases} \qquad (II.46)$$

It's easy to derive :

$$A = \frac{T_e - T_i}{\ln\left(\frac{r_e}{r_i}\right)} \text{ et } B = \frac{T_i \ln r_e - T_e \ln r_i}{\ln\left(\frac{r_e}{r_i}\right)}$$

By replacing A and B in equation (II. 45) :

$$T(r) = \frac{T_e - T_i}{\ln\left(\frac{r_e}{r_i}\right)}\ln r + \frac{T_i \ln r_e - T_e \ln r_i}{\ln\left(\frac{r_e}{r_i}\right)} \qquad (II.47)$$

or :

$$T(r) = T_i + \frac{T_i - T_e}{\ln\left(\frac{r_i}{r_e}\right)}\ln\left(\frac{r}{r_i}\right) \qquad (II.48)$$

Given Fourier's law :

$$\varphi = -\lambda \frac{dT(r)}{dr}$$

For a length L of the cylinder, the expression for the heat flux density is :

$$\varphi = -\lambda \frac{T_i - T_e}{\ln\left(\frac{r_i}{r_e}\right)} \frac{1}{r} \qquad (II.49)$$

The expression for the heat flux passing through each cylindrical surface is then :

$$\Phi = \varphi S = \lambda\, 2\pi L \frac{T_i - T_e}{\ln\left(\frac{r_e}{r_i}\right)} \qquad (II.50)$$

hence the expression for the thermal resistance of the hollow cylinder :

$$R = \frac{\ln\left(\frac{r_e}{r_i}\right)}{\lambda\, 2\pi L} \qquad (II.51)$$

In the particular case of a thin tube of thickness e, and using the approximation $\ln(1 + \frac{e}{r_i}) \approx \frac{e}{r_i}$, the above results simplify as follows:

$$\varphi = \lambda \frac{T_i - T_e}{e} \frac{r_i}{r} \; ; \; \Phi = \lambda\, 2\pi r_i L \frac{T_i - T_e}{e} \; ; \; R = \frac{e}{\lambda\, 2\pi r_i L}$$

II.3.5 Critical insulation thickness

Insulating heating pipes involves thermally insulating these pipes in order to limit heat loss and improve the energy efficiency of the heating distribution circuit.

Consider a layer of insulation, of thermal conductivity λ, that could be installed around a circular pipe. The inside temperature of the insulation is T_i on *ri* and the outside surface is exposed to a convection environment at *T*", as shown in Figure II.9. In this case, the heat flux through the insulation is given by equation :

$$\Phi = \frac{2\pi L\,(T_i - T_\infty)}{\frac{\ln\left(r_e/r_i\right)}{\lambda} + \frac{1}{r_e\,h}} \qquad \text{(II.52)}$$

Let's now manipulate this expression to determine the outer radius of the insulation *re* that will maximise heat transfer. The maximisation condition is :

$$\frac{d\Phi}{dr_e} = 0 = \frac{-2\pi L\,(T_i - T_\infty)\left(\frac{1}{\lambda\,r_e} - \frac{1}{h\,r_e^2}\right)}{\frac{\ln\left(r_e/r_i\right)}{\lambda} + \frac{1}{r_e\,h}}$$

which gives the result :

$$r_e = \frac{\lambda}{h} \qquad \text{(II.53)}$$

Equation (II.53) expresses the concept of the critical insulation radius. This means that, for this value of *re*, the curve of variation in total thermal resistance as a function of insulation thickness passes through a minimum. As a result, increasing the thickness of the insulation layer means that the pipe is less well insulated! Physically, this can be explained as follows: if *re*
increases, the external exchange surface also increases, which tends to reduce the effect of the insulation.

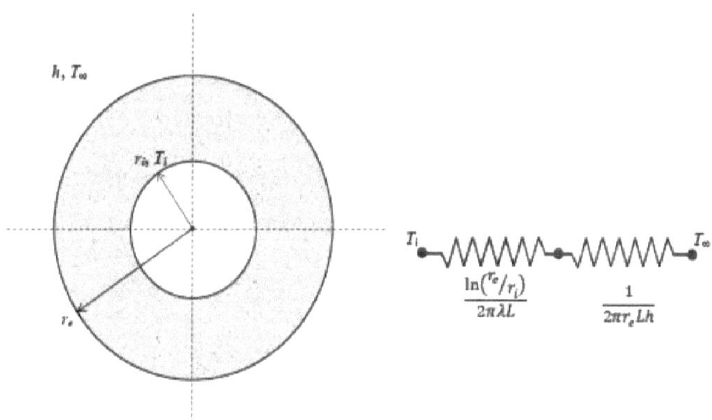

FIGURE II.9: Critical insulation thickness.

II.3.6 Hollow sphere

This problem is analogous to that of the hollow cylinder. A homogeneous hollow sphere of inner radius *ri* and outer radius *re*. Its boundary surfaces, at *ri* and *re*, are maintained at uniform and constant temperatures *Ti* and *Te* (Fig. II.10). The isotherms are concentric spherical surfaces and the temperature is a function only of the radius *r*. Since the heat flow is radial, the equation for *T* in spherical coordinates is written :

$$\frac{d^2T}{dr^2}+\frac{2}{r}\frac{dT}{dr}=0 \qquad \text{(II.54)}$$

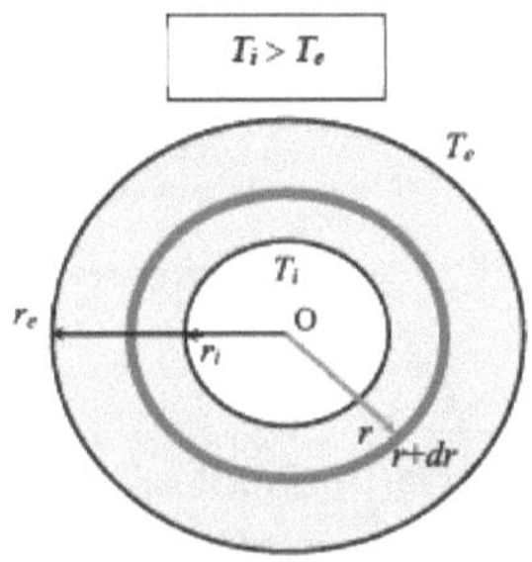

FIGURE II.10: Conduction in a hollow sphere.

Equation (II.54) can also be written as :

$$\frac{1}{r^2}\frac{d}{dr}\left(r^2\frac{dT}{dr}\right)=0 \qquad \text{(II.55)}$$

a first integration gives :

$$r^2\frac{dT}{dr}=\text{A} \Rightarrow \frac{dT}{dr}=\frac{\text{A}}{r^2}$$

leading to the solution of equation (II.55) such that :

$$T(r)=-\frac{\text{A}}{r}+\text{B} \qquad \text{(II.56)}$$

Consider Dirichlet-type boundary conditions and determine the constants A and B :

$$\begin{cases} T=T_e \text{ si } r=r_e \\ T=T_i \text{ si } r=r_i \end{cases}$$

We obtain the system :

$$\begin{cases} T_e=-\dfrac{A}{r_e}+B \\ T_i=-\dfrac{A}{r_i}+B \end{cases} \qquad \text{(II.57)}$$

from which we can easily derive :

$$A = \frac{T_e - T_i}{\left(\frac{1}{r_i} - \frac{1}{r_e}\right)} \text{ et } B = T_i + \frac{T_e - T_i}{\left(\frac{1}{r_i} - \frac{1}{r_e}\right)} \frac{1}{r_i}$$

By replacing A and B in equation (II.56), the equation giving the radial temperature distribution in a hollow sphere :

$$T(r) = T_i - (T_i - T_e) \frac{\frac{1}{r_i} - \frac{1}{r}}{\frac{1}{r_i} - \frac{1}{r_e}} \qquad \text{(II. 58)}$$

11.4 Heat equation

11.4.1 Energy balance

We will apply the law of conservation of energy (first principle of thermodynamics) to a finite volume *v* of surface area *s* contained in a medium of total volume V (Fig. II.11).

In what follows, we will assume that the medium under consideration is homogeneous and isotropic and that the volume *v* contains internal sources giving off heat with a power density *pv* (by the Joule effect, for example). Thermal characteristics such as mass heat *C*, density *p* and thermal conductivity *λ* can be a function of the point under consideration and the temperature at that point. In addition, *pv* can be a function of time.

The amount of heat exchanged with the outside world per unit time comprises the power exchanged at the boundary surface *s* and the power supplied by the sources. The power passing through *s* in the direction of the exterior normal $\vec{n}$ was described earlier (Eq. II.3).

The algebraic power received by *v*, which passes through *s* in the opposite direction to n, has the expression :

$$\Phi = \iint -(\vec{\varphi} \cdot \vec{n})\, ds \qquad \text{(II.59)}$$

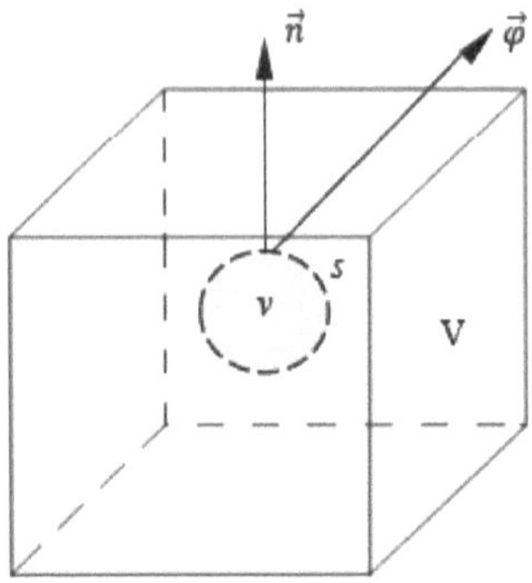

FIGURE II.11: Heat balance.

The power generated per unit volume (W/m$^{3)}$, due to internal heat sources, is given by the integral :

$$\Phi_v = \iiint p_v\, dv \qquad \text{(II.60)}$$

As the material system under consideration is an undeformable solid, only the local variation in temperature is taken into account in the expression of the instantaneous variation in internal energy, i.e. :

$$\Delta U = \iiint \rho\, C\, \frac{\partial T}{\partial t}\, dv \qquad \text{(II.61)}$$

The first principle of thermodynamics can then be expressed as :

$$\Delta U = \Phi + \Phi_v$$

i.e. :

$$\iiint \rho\, C\, \frac{\partial T}{\partial t}\, dv = \iint -(\vec{\varphi} \cdot \vec{n})\, ds + \iiint p_v\, dv \qquad \text{(II.62)}$$

Ostrogradsky's formula can be used to transform the surface integral into a volume integral:

$$\iint (\vec{\varphi} \cdot \vec{n})\, ds = \iiint div\, \vec{\varphi}\, dv \qquad \text{(II.63)}$$

Thus the relationship (II.62) becomes :

$$\iiint \left[- div\, \vec{\varphi} + p - \rho\, C\, \frac{\partial T}{\partial t}\right] dv = 0 \qquad \text{(II.64)}$$

v being an arbitrary volume of V, the instantaneous heat balance at each point of the total volume is expressed as :

$$div\,\vec{\varphi} + \rho\,C\,\frac{\partial T}{\partial t} - p_v = 0 \qquad (II.65)$$

By replacing the flux density by its expression given by Fourier's law, the heat equation is written :

$$-\,div\left(\lambda(T)\,\overrightarrow{grad}\,T\right) + \rho\,C\,\frac{\partial T}{\partial t} - p_v = 0 \qquad (II.66)$$

Noting :

$$div\left(\lambda(T)\,\overrightarrow{grad}\,T\right) = \lambda(T)\,div\,\overrightarrow{grad}\,T + \overrightarrow{grad}\,T\,\overrightarrow{grad}\,\lambda(T)$$

This gives the non-linear heat equation :

$$\nabla^2 T - \frac{1}{\alpha}\frac{\partial T}{\partial t} + \frac{1}{\lambda}\frac{d\lambda}{dT}\left(\overrightarrow{grad}\,T\right)^2 + \frac{p_v}{\lambda} = 0 \qquad (II.67)$$

where V^2denotes the Laplacian (also denoted Δ) and *a* the thermal diffusivity such that :

$$\alpha = \frac{\lambda}{\rho\,C}$$

The thermal diffusivity of a solid is expressed in m^2/s, like the kinematic viscosity of a fluid.

11.4.2 Forms of the heat equation

In order to linearise equation (II.67), we must assume a medium with constant thermal characteristics *(Л, p* and *C)* and for which *pv* is independent of temperature. In such a situation, the classical linear heat equation is :

$$\nabla^2 T - \frac{1}{\alpha}\frac{\partial T}{\partial t} + \frac{p_v}{\lambda} = 0 \qquad (II.68)$$

Depending on the nature of the problem under consideration, the heat equation takes different forms. The usual equations to be solved, which are very frequently encountered in heat transfer problems, are easily deduced from equation (II.68):

- Laplace's equation (medium without internal sources in steady state),

$$\nabla^2 T = 0 \qquad (II.69)$$

- Poisson's equation (medium with internal sources in steady state),

$$\nabla^2 T + \frac{p_v}{\lambda} = 0 \qquad (II.70)$$

- Fourier equation (medium without internal sources in transient regime),

$$\nabla^2 T - \frac{1}{\alpha}\frac{\partial T}{\partial t} = 0 \qquad (II.71)$$

The forms of the V2 operator of the temperature function *T* in the different Cartesian, cylindrical and spherical coordinate systems respectively are :

$$\nabla^2 T = \frac{\partial^2 T}{\partial x^2} + \frac{\partial^2 T}{\partial y^2} + \frac{\partial^2 T}{\partial z^2}$$

$$\nabla^2 T = \frac{\partial^2 T}{\partial r^2} + \frac{1}{r}\frac{\partial T}{\partial r} = \frac{1}{r}\frac{\partial}{\partial r}\left(r\frac{\partial T}{\partial r}\right)$$

$$\nabla^2 T = \frac{\partial^2 T}{\partial r^2} + \frac{2}{r}\frac{\partial T}{\partial r}$$

11.4.3 Space-time boundary conditions

The general heat equation (II.68) expresses, by means of a relationship between the variables *x*, *y*, *z*, *t* and the temperature *T*, the mechanism of heat transmission by conduction at any point *M* (*x*, *y*, *z*) and at any instant *t*. In principle, this second-order linear partial differential equation has an infinite number of solutions, involving arbitrary constants or functions. However, of these mathematical solutions, only one will make any physical sense, once the thermophysical problem under study has been correctly posed, i.e. :

- Define the geometry of the solid body.
- Know the initial temperature distribution, inside the solid and on its surface, at time $t = 0$.
- Express the real physical conditions at the surface of the solid in stylised mathematical form.

These conditions are the causes that determine the subsequent evolution of the phenomenon. Since the heat equation describes an irreversible phenomenon, this implies that the medium under consideration has been subjected, at the initial instant, to a break in its thermal equilibrium.

It is therefore necessary to know the initial temperature distribution and/or the normal derivative of the *T* function on the boundary surface.

Convergence when solving the heat equation, which in principle admits an infinite number of solutions, depends essentially on the boundary conditions, in addition, of course, to geometric, physical and temporal considerations. The boundary conditions describe the particular features of the process at the boundary surface, i.e. at the body wall, for $t > 0$. The most frequently encountered spatial boundary conditions are as follows:

- If the wall temperature at the surface *Ts* is a constant, we speak of an isothermal condition. This can be an imposed temperature, the 1st type (or Dirichlet) condition:

$$T_s = f(P_s, t) = f(x_s, y_s, z_s, t)$$

Ps(xs, ys, zs) being a point on the surface.

In the case where the heat flux density is imposed, this is a boundary condition of the 2nd (or Neumann) type: $\varphi = -\lambda \left(\frac{\partial T}{\partial n}\right)_s = f(P_s, t)$; derived normal to the surface.

If the body is thermally isolated, the flux is zero at any point on its surface. It is said to be an adiabatic body, i.e. :

$$\left(\frac{\partial T}{\partial n}\right)_s = 0$$

If the body is heated or cooled by a fluid, heat transfer takes place by convection. The flux density across the boundary surface is proportional to the temperature difference between the wall and the surrounding fluid. This is the 3rd type (or Fourier) boundary condition. We use the Newton-Riemann hypothesis for mixed heat transfer, where :

$$-\lambda \left(\frac{\partial T}{\partial n}\right)_s = \varphi = h(T_s - T_\infty)$$

h is the convective exchange coefficient and *T*" is the temperature of the fluid away from the wall. This is the condition for heat transfer by convection, known as Newton's law.

Hypothetically, the solid body in question can exchange a certain amount of heat with its external environment by radiation. The heat flux density exchanged by radiation is not studied in this book. However, it is important to realise that this mode of transfer by radiation may be an important surface condition to express, in order to be able to solve the conduction problem.

Generally speaking, the different types of surface conditions can be superimposed. For example, heat passing through a metal plate by conduction is transmitted to the surrounding environment by both convection and radiation.

11.5 Solving the heat equation

II.5.1 Steady state

In steady state, the temperature at each point in the medium is independent of time. The general equation can be reduced to Laplace's equation (II.69) for a medium without sources, and to Poisson's equation (II.70) when the medium has internal sources.

As soon as the unidirectional case is abandoned, any attempt at an analytical solution quickly becomes inextricable. In fact, theoretical

calculations, which were the only ones available for many years, have given way to numerical methods, which now hold a virtual monopoly. Nowadays, specialised calculation codes are available on microcomputers in user-friendly versions.
The principle of numerical solution methods applied in calculation codes is to transform the heat equation and the boundary conditions into a system of finite difference or finite element equations. In order to better appreciate the practical significance of the numerical method, in what follows we will study a simple heat transfer problem in a two-dimensional domain.
The most general form of the heat equation, with no internal heat sources and steady state conditions, is given below:

$$\frac{\partial^2 T}{\partial x^2}+\frac{\partial^2 T}{\partial y^2}+\frac{\partial^2 T}{\partial z^2}=0 \qquad (II.72)$$

11.5.1.1 2D plate

The general concept of the numerical method applied is to divide the domain under study into regions at the centre of which a node is placed, and to carry out a heat balance for each node, taking into account the temperature of the node in question, those of neighbouring nodes and, of course, the thermal properties of the material. Of course, we can also transform the differential equations directly into finite difference equations without performing a heat balance.
We create heat transfer in a square metal plate measuring (0.5 x 0.5) m^2, by imposing a temperature of 0°C along its sides AB and AD, and a temperature of 100°C at its corner C
(Fig. II.12). The temperature distribution $T(x, y)$ in the plate is stationary. It is a solution of Laplace's equation:

$$\frac{\partial^2 T}{\partial x^2}+\frac{\partial^2 T}{\partial y^2}=0 \qquad (II.73)$$

A good command of the basics of numerical analysis is essential to solve the above equation:

- The discretisation of a function $f(x)$.
- An approximation of the first derivative $f(x)$.
- An approximation of the second derivative $f''(x)$.
- The discretisation of a function $f(x, y)$.
- An approximation of the partial derivatives.

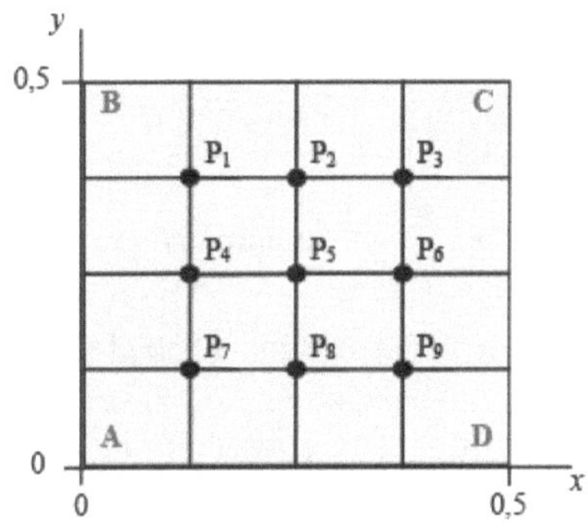

FIGURE II.11: Temperature distribution in a 2D plate.

The Taylor series expansion of the function $T(x', y)$ around a point with coordinates (x, y) is written :

$$T(x+\Delta x, y) = T(x,y) + \Delta x \frac{\partial T}{\partial x} + \frac{(\Delta x)^2}{2!}\frac{\partial^2 T}{\partial x^2} + \frac{(\Delta x)^3}{3!}\frac{\partial^3 T}{\partial x^3} + \cdots$$

$$T(x-\Delta x, y) = T(x,y) - \Delta x \frac{\partial T}{\partial x} + \frac{(\Delta x)^2}{2!}\frac{\partial^2 T}{\partial x^2} - \frac{(\Delta x)^3}{3!}\frac{\partial^3 T}{\partial x^3} + \cdots$$

Adding the two relations above and dividing by $(\Delta x)^2$ we find:

$$\frac{\partial^2 T}{\partial x^2} = \frac{T(x+\Delta x, y) - 2\,T(x,y) + T(x-\Delta x, y)}{(\Delta x)^2} + 0(\Delta x)^2$$

Operating in the same way for the variable y we find :

$$\frac{\partial^2 T}{\partial y^2} = \frac{T(x, y+\Delta y) - 2\,T(x,y) + T(x, y-\Delta y)}{(\Delta y)^2} + 0(\Delta y)^2$$

Positing $\Delta x = \Delta y = h$, the two-dimensional Laplacian is then written:

$$\frac{\partial^2 T}{\partial x^2} + \frac{\partial^2 T}{\partial y^2} = \frac{T(x+\Delta x, y) + T(x-\Delta x, y) - 4\,T(x,y)}{h^2} +$$

$$+ \frac{T(x, y+\Delta y) + T(x, y-\Delta y)}{h^2} + 0(\Delta x)^2 + 0(\Delta y)^2$$

Given the nodal representation in Figure (II.13), we can write the previous expression in a simpler form such that:

$$\frac{\partial^2 T}{\partial x^2} + \frac{\partial^2 T}{\partial y^2} = \frac{T_{i+1,j} + T_{i-1,j} - 4T_{i,j} + T_{i,j+1} + T_{i,j-1}}{h^2} = 0$$

For each internal node, the discretised Laplace equation gives an approximation of the temperature $T(x, y)$, which enables it to be solved numerically:

$$T_{i,j} = \frac{T_{i+1,j} + T_{i-1,j} + T_{i,j+1} + T_{i,j-1}}{4} \qquad \text{(II.74)}$$

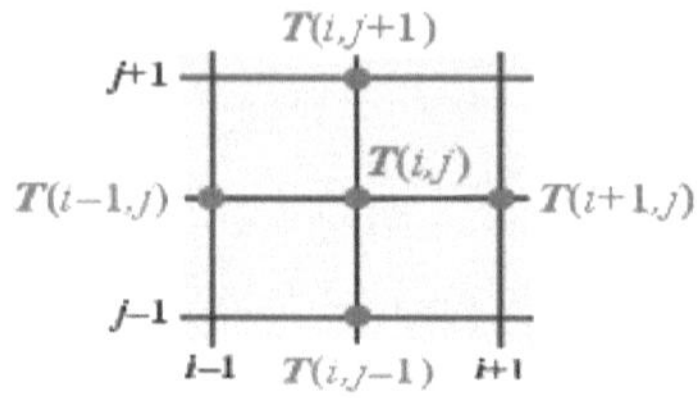

FIGURE II.13: Nodal representation of temperature.

The main numerical methods for solving fluid mechanics and thermal engineering problems incorporated in simulation codes are: finite differences, finite elements and finite volumes. These methods are either programmed using a specific computer language, or are already programmed. This is the case with commercial codes such as COMSOL Multiphysics, CFD Fluent, ANSYS, etc. Nevertheless, mastery and in-depth knowledge will enable these digital tools to be used intelligently.

As an example, we are going to apply a finite-difference calculation code, available on a microcomputer in a stationary 2D version, to the solution of the previous problem of the plate with imposed wall temperatures.

Figures (II.14) and (II.15) show, respectively, the mesh used, consisting of 200 elements and 121 internal nodes, and a temperature map divided into ten bands of values.

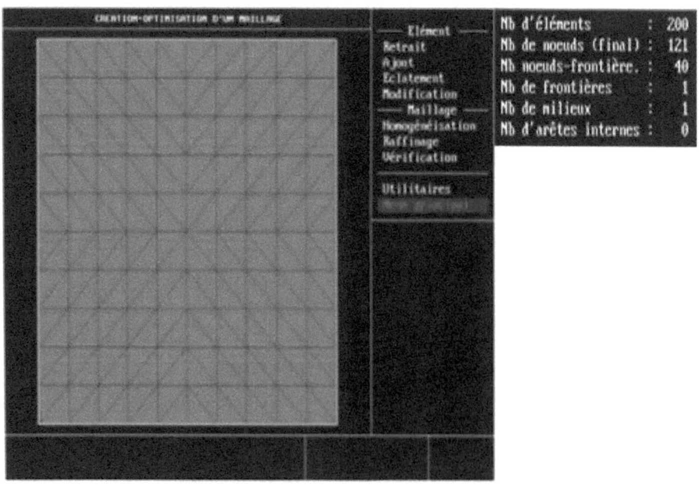

FIGURE II.14: Mesh of the plate.

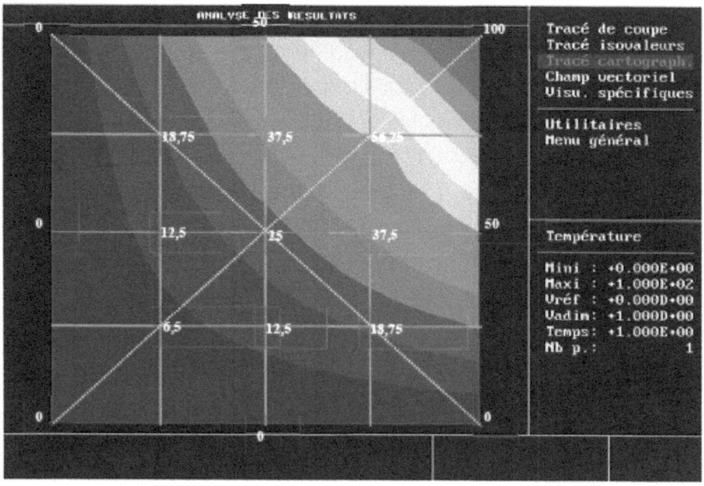

FIGURE II.15: Visualisation of isotherms in the plate.

11.5.1.2 Square section furnace

We propose to study the temperature distribution in the walls of a square-section furnace (Fig. II.16). We denote *Ti* the wall temperature at the inner boundary (Γi*)* of the furnace and *Te* the wall temperature at the outer boundary of the furnace (*Γe*).

The temperature *T*(*x, y*) at any point *M*(*x, y*) on the wall satisfies Laplace's equation, with boundary conditions :

$$\begin{cases} \Gamma_i \rightarrow T(x,y) = T_i \\ \Gamma_e \rightarrow T(x,y) = T_e \end{cases}$$

Assuming :

$$\begin{cases} x_i = i.\Delta x = i.h, 1 \leq i \leq n \\ y_j = j.\Delta y = j.h, 1 \leq j \leq m \end{cases}$$

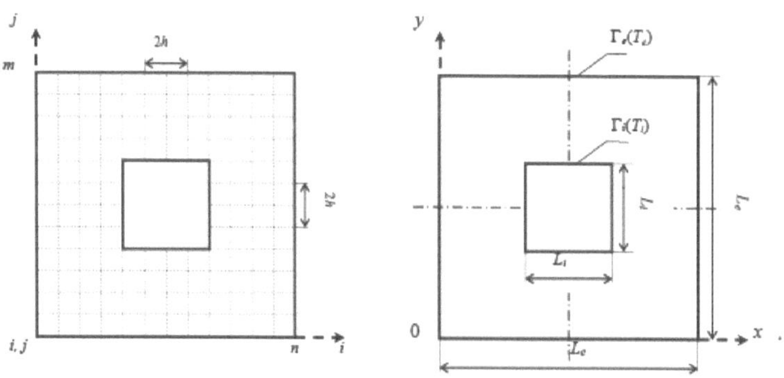

FIGURE II.16: Oven with square cross-section.

In steady state (or stationary) conditions, the temperature field is independent of time and the isothermal surfaces are fixed. This is a heat conduction problem with Dirichlet boundary conditions.

Because of the symmetry of the problem and the conservation of heat flow, it will suffice to study a quarter of the cross-section of the furnace (Fig. II.17). For i = 7 and j = 7 (in this case n = m = 13), we obtain 27 internal nodes. The temperatures at the boundaries Γ_i and Γ_e of dimensions Li = 200 mm and Le = 600 mm are 1220 °C and 35 °C respectively.

This involves solving a system of 27 equations with as many unknowns for the following boundary conditions:

$$\begin{cases} T_{5,5} = T_{5,6} = T_{5,7} = T_{6,5} = T_{7,5} = T_i = 1220\,°C \\ T_{1,1} = T_{1,2} = T_{2,1} = \cdots = T_{1,7} = T_{7,1} = T_e = 35\,°C \end{cases}$$

Because of the symmetry of the problem :

$$\begin{cases} T_{2,6} = T_{2,8}, T_{3,6} = T_{3,8}, T_{4,6} = T_{4,8} \\ T_{6,2} = T_{8,2}, T_{6,3} = T_{8,3}, T_{6,4} = T_{8,4} \end{cases}$$

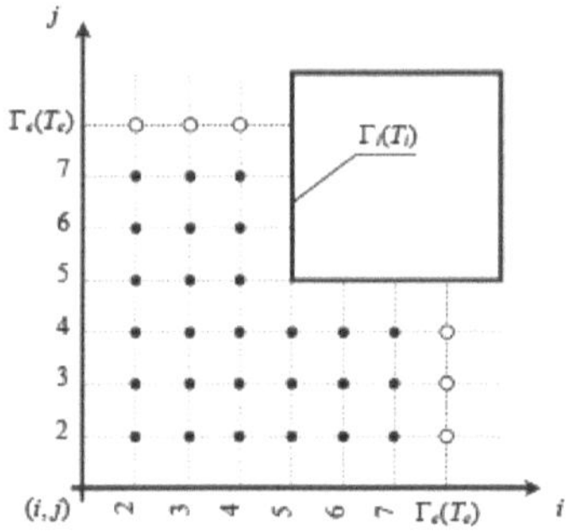

FIGURE II.17: Mesh of the quarter of the straight section of the furnace.

Each node (black dots in Figure II.17) represents an unknown temperature which must be determined by solving the linear matrix system below.

There are several numerical methods for solving systems of linear equations. A direct method of solving a linear system is an algorithm which, if the computer made exact calculations, would give the solution in a finite number of operations. There are also iterative methods, which consist of constructing a sequence of vectors converging towards the vector x, the solution to the system $Ax = b$. The advantage of iterative methods, compared with direct methods, is that they are simple to program and require less memory space. On the other hand, they often take longer to calculate.

The Gauss-Seidel algorithm, which is generally faster during iteration, is preferred for matrix solving. In addition, convergence is guaranteed whether the matrix A is a matrix with a strict dominant diagonal by rows or a symmetric positive-definite matrix. The results can be executed and displayed using MATLAB, a language suitable for matrix calculations. The temperature fields are visualised in a three-dimensional environment and in parallel planes (Fig. II.18).

The mesh size influences the accuracy, convergence and speed of a simulation. A finer (or dense) mesh will generally give a more accurate solution. However, a high mesh density requires a longer calculation time and more RAM on the computer used. ANSYS multiphysics simulation software provides tools to produce the most appropriate mesh for accurate solutions.

$$\left[\begin{array}{ccccccccccccccccccccccccccc}
-4&1&0&0&0&0&1&0&0&0&0&0&0&0&0&0&0&0&0&0&0&0&0&0&0&0&0\\
1&-4&1&0&0&0&0&1&0&0&0&0&0&0&0&0&0&0&0&0&0&0&0&0&0&0&0\\
0&1&-4&1&0&0&0&0&1&0&0&0&0&0&0&0&0&0&0&0&0&0&0&0&0&0&0\\
0&0&1&-4&1&0&0&0&0&1&0&0&0&0&0&0&0&0&0&0&0&0&0&0&0&0&0\\
0&0&0&1&-4&1&0&0&0&0&1&0&0&0&0&0&0&0&0&0&0&0&0&0&0&0&0\\
0&0&0&0&2&-4&0&0&0&0&0&1&0&0&0&0&0&0&0&0&0&0&0&0&0&0&0\\
1&0&0&0&0&0&-4&1&0&0&0&0&1&0&0&0&0&0&0&0&0&0&0&0&0&0&0\\
0&1&0&0&0&0&1&-4&1&0&0&0&0&1&0&0&0&0&0&0&0&0&0&0&0&0&0\\
0&0&1&0&0&0&0&1&-4&1&0&0&0&0&1&0&0&0&0&0&0&0&0&0&0&0&0\\
0&0&0&1&0&0&0&0&1&-4&1&0&0&0&0&1&0&0&0&0&0&0&0&0&0&0&0\\
0&0&0&0&1&0&0&0&0&1&-4&1&0&0&0&0&1&0&0&0&0&0&0&0&0&0&0\\
0&0&0&0&0&1&0&0&0&0&2&-4&0&0&0&0&0&1&0&0&0&0&0&0&0&0&0\\
0&0&0&0&0&0&1&0&0&0&0&0&-4&1&0&0&0&0&1&0&0&0&0&0&0&0&0\\
0&0&0&0&0&0&0&1&0&0&0&0&1&-4&1&0&0&0&0&1&0&0&0&0&0&0&0\\
0&0&0&0&0&0&0&0&1&0&0&0&0&1&-4&1&0&0&0&0&1&0&0&0&0&0&0\\
0&0&0&0&0&0&0&0&0&1&0&0&0&0&1&-4&1&0&0&0&0&0&0&0&0&0&0\\
0&0&0&0&0&0&0&0&0&0&1&0&0&0&0&1&-4&1&0&0&0&0&0&0&0&0&0\\
0&0&0&0&0&0&0&0&0&0&0&1&0&0&0&0&2&-4&0&0&0&0&0&0&0&0&0\\
0&0&0&0&0&0&0&0&0&0&0&0&1&0&0&0&0&0&-4&1&0&1&0&0&0&0&0\\
0&0&0&0&0&0&0&0&0&0&0&0&0&1&0&0&0&0&1&-4&1&0&1&0&0&0&0\\
0&0&0&0&0&0&0&0&0&0&0&0&0&0&1&0&0&0&0&1&-4&0&0&1&0&0&0\\
0&0&0&0&0&0&0&0&0&0&0&0&0&0&0&0&0&0&1&0&0&-4&1&0&1&0&0\\
0&0&0&0&0&0&0&0&0&0&0&0&0&0&0&0&0&0&0&1&0&1&-4&1&0&1&0\\
0&0&0&0&0&0&0&0&0&0&0&0&0&0&0&0&0&0&0&0&1&0&1&-4&0&0&1\\
0&0&0&0&0&0&0&0&0&0&0&0&0&0&0&0&0&0&0&0&0&2&0&0&-4&1&0\\
0&0&0&0&0&0&0&0&0&0&0&0&0&0&0&0&0&0&0&0&0&0&2&0&0&-4&1\\
0&0&0&0&0&0&0&0&0&0&0&0&0&0&0&0&0&0&0&0&0&0&0&2&0&1&-4
\end{array}\right]$$

$$
\times
\begin{bmatrix}
T_{2,2}\\ T_{3,2}\\ T_{4,2}\\ T_{5,2}\\ T_{6,2}\\ T_{7,2}\\ T_{2,3}\\ T_{3,3}\\ T_{4,3}\\ T_{5,3}\\ T_{6,3}\\ T_{7,3}\\ T_{2,4}\\ T_{3,4}\\ T_{4,4}\\ T_{5,4}\\ T_{6,4}\\ T_{7,4}\\ T_{2,5}\\ T_{3,5}\\ T_{4,5}\\ T_{2,6}\\ T_{3,6}\\ T_{4,6}\\ T_{2,7}\\ T_{3,7}\\ T_{4,7}
\end{bmatrix}
=
\begin{bmatrix}
-2T_e\\ -T_e\\ -T_e\\ -T_e\\ -T_e\\ -T_e\\ -T_e\\ 0\\ 0\\ 0\\ 0\\ 0\\ -T_e\\ 0\\ 0\\ -T_i\\ -T_i\\ -T_i\\ -T_e\\ 0\\ -T_i\\ -T_e\\ 0\\ -T_i\\ -T_e\\ 0\\ -T_i
\end{bmatrix}
$$

By way of comparison, a visualisation of the previous results is carried out in ANSYS in order to highlight the impact of a dense mesh on the thermal map of the temperature distribution (Fig. II.19).

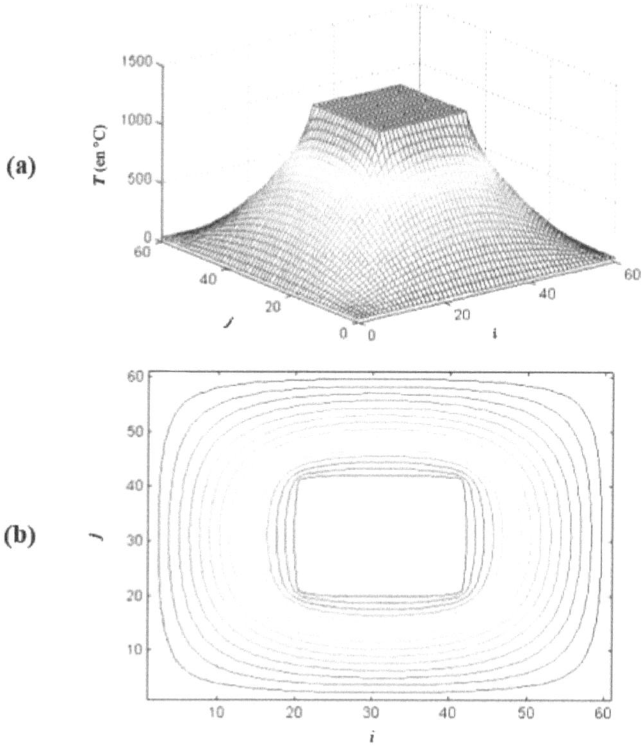

<u>Figure II.18</u>: Thermal simulation of the furnace wall using MATLAB: (a) Temperature profile, (b) Isotherms.

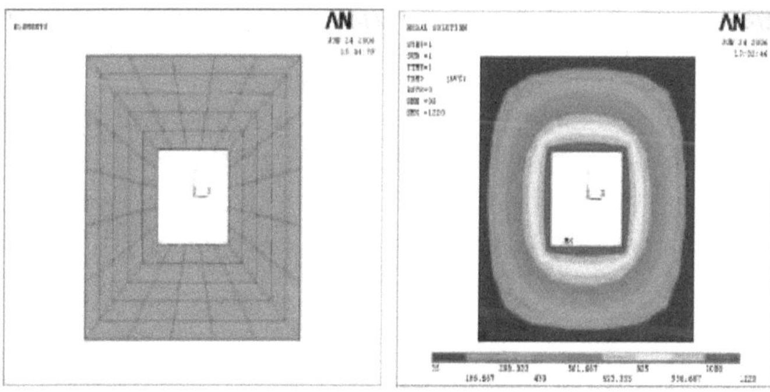

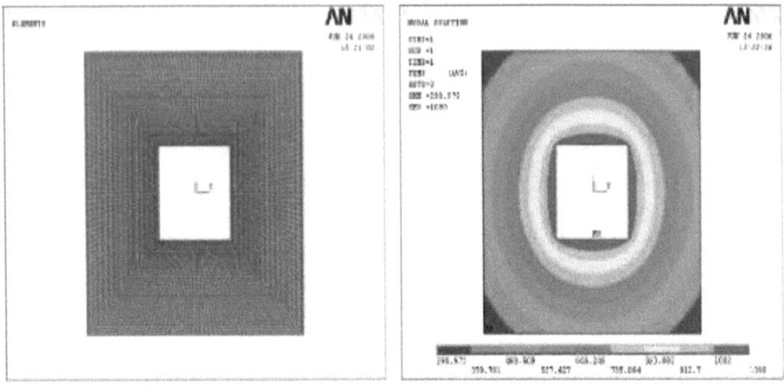

Figure II.19: Thermal simulation of the furnace wall using ANSYS:
Influence of the mesh size on the accuracy of the results.

11.5.1.3 Bar problem

In order to illustrate in a concrete way the methods for numerically solving heat conduction problems, we are going to deal entirely with a simple case which is nonetheless impossible to solve using a non-numerical method. This is the typical problem of conduction in a solid, with lateral exchange towards the external medium, generally consisting of a fluid at rest (exchange by natural convection) or in motion (exchange by forced convection).

Consider a bar of constant cross-section and length much greater than its transverse dimensions (Fig. II.20), made of a homogeneous material which is a good conductor of heat. The bar is heated continuously at both ends, which means that end A is kept at a constant temperature T_A and end B at a constant temperature T_B.

The heat propagates along the axis of the bar, and radially, to flow into the ambient medium, of temperature T'', through the lateral surface of the bar. Assuming steady-state equilibrium is reached, we propose to calculate the temperature distribution along the bar.

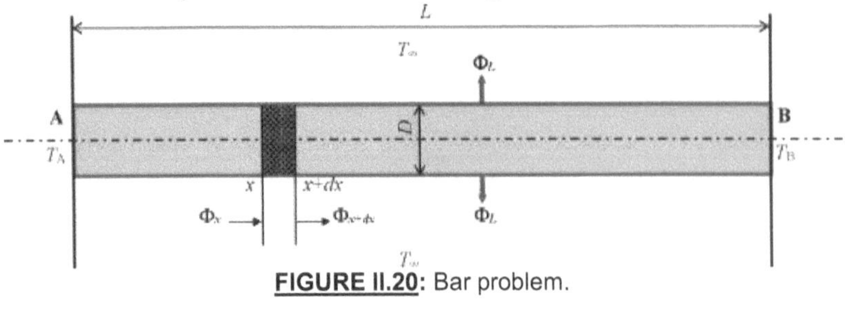

FIGURE II.20: Bar problem.

11.5.1.3.1 Modelling the physical problem

We will specify the problem by considering a cylindrical bar, of diameter *D* and length *L* much greater than *D*. The radial flux density is :

$$\varphi_r = -\lambda \frac{\partial T}{\partial r}$$

Since the bar is made of a material that is a good conductor of heat, radial dissipation requires only small gradients. $\frac{\partial T}{\partial r}$.

Since the transverse dimensions are small, the transverse variations in temperature will be negligible and a straight section with abscissa *x* can be considered isothermal. The problem has therefore become one-dimensional, the object of this problem being to calculate the temperature distribution *T*(*x*) along the bar.

We are only interested in the steady state. The thermal equilibrium that characterises this steady state necessarily results in a zero algebraic sum of the incoming and outgoing heat fluxes of any section of the bar, between the abscissas *x* and *x* + dx.

Let us denote by *Φx* the heat flux passing through the straight section of area *S* and abscissa *x* and by *ΦL* the heat flux exchanged by the lateral surface (*πD dx*) with the ambient air.

The energy balance of the slice is zero in steady state, i.e. :

$$\Phi_x - \Phi_{x+dx} - \Phi_L = 0 \qquad (II.75)$$

Φx is given by Fourier's law:

$$\Phi_x = -\lambda \left[\frac{dT}{dx}\right]_x$$

ΦL is the heat flux exchanged by the lateral surface. To calculate it, we need to introduce a convective heat exchange coefficient.

On the surface of a solid body, consider an element of area *dS* at temperature *Tp* (Fig. II.21). If the body is in contact with a moving fluid medium characterised by a temperature *T*", the quantity of heat passing through *dS* during the time interval *dt* can be written as :

$$d^2Q = h\left(T_p - T_\infty\right) dS\, dt \qquad (II.76)$$

The coefficient *h* is the convective thermal conductance, or convective heat exchange coefficient. *h* is expressed in W/(m$^{(2)}$·K).

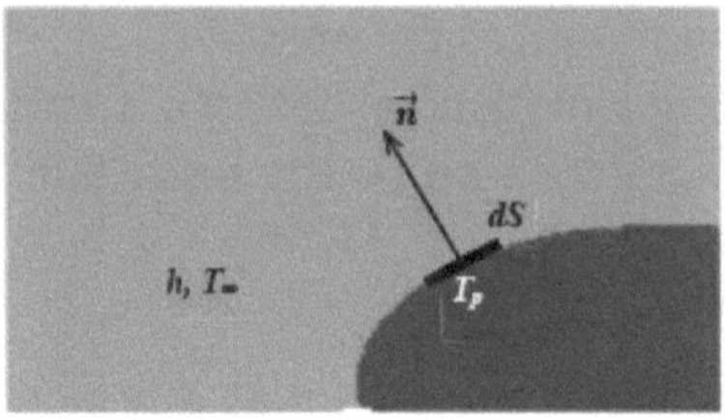

FIGURE II.21: Mixed heat transfer.

Any calculation of heat exchange by convection requires the determination of the coefficient h, which is often a complicated task, as heat transfer by convection is a complex phenomenon.

For gases, the natural convection transfer coefficients h are always of the order of 1 to 10 $W/(m^{(2)} \cdot K)$. We will therefore limit ourselves to a summary diagram to take account of convective transfer, which increases the further away we are from the two heated ends. We will assume a minimum value equal to 2 $W/(m^{(2)} \cdot K)$ at A and B and a maximum value twice that at the middle of the bar, hence the following simple model for the transfer coefficient $h(x)$, in which we have adopted a parabolic law of variation:

$$h(x) = 2 - 8x(x-1)$$

For Φ_L, therefore, we adopt the expression :

$$\Phi_L = h(x)[T(x) - T_\infty]\,\pi\, D\, dx$$

Finally, the energy balance for a slice of thickness dx can be written as :

$$\lambda\left\{\left[\frac{dT}{dx}\right]_{x+dx} - \left[\frac{dT}{dx}\right]_x\right\}\frac{\pi D^2}{4} - h(x)[T(x) - T_\infty]\,\pi\, D\, dx = 0 \qquad \text{(II.77)}$$

Or :

$$\frac{d^2T}{dx^2} - (T - T_\infty)\eta^2(x) = 0 \qquad \text{(II.78)}$$

with :

$$\eta^2(x) = \frac{4h(x)}{\lambda\, D}$$

η, which has the inverse dimension of a length, characterises the speed with which the heating imparted to the bar of constant circular cross-section fades along its length.

11.5.1.3.2 Discretisation of the physical problem

The bar AB under consideration has a length L, which will be represented by the segment [0,1] of the O x axis. The segment [0,1] is divided into N+1 intervals of length $\Delta x = \frac{1}{N+1}$.

The N+2 necessary subdivision points (including the ends A and B), or nodes *xi* of this mesh, will then be defined as follows:

$$x_i = i\Delta x, \quad i \in \{0, \ldots, N+1\}$$

The discrete problem consists in finding an ft approximation of the temperature $T(x_i)$ at each of the internal nodes (i.e. x_i, $1 \leq i \leq N$) of the mesh, the solution being imposed at the $x_0 = 0$ et $x_{N+1} = 1$ bounds of the interval. These different values θ are solutions to the discrete problem :

$$-\frac{1}{(\Delta x)^2}[\theta_{i+1} - 2\theta_i + \theta_{i-1}] + \eta^2(x_i)\theta_i = T_\infty \eta^2(x_i), \quad 1 \leq i \leq N \quad \text{(II.79)}$$

where : $\theta_0 = T_A$, $\theta_{N+1} = T_B$.

This finite difference scheme replaces the physical problem previously modelled by the following equations:

$$-\frac{d^2T(x)}{dx^2} + \eta^2(x)T(x) = T_\infty \eta^2(x), \quad 0 < x < 1 \qquad \text{(II.80)}$$

where : $T(0) = T_A$, $T(1) = T_B$.

11.5.1.3.3 Solving algorithm

By reordering the terms of equation (II.79), a linear system of *N* equations with *N* unknowns is revealed:

$$\frac{1}{(\Delta x)^2}\{-\theta_{i+1} + [2 + (\Delta x)^2\eta^2(x_i)]\theta_i - \theta_{i-1}\} = T_\infty \eta^2(x_i) \quad \text{(II.81)}$$

$i = 1, N$

The first equation in the system considers the condition imposed at the end A of the bar. By making *I* = 1, we show the influence exerted on the first internal node by the end condition $\theta_0 = T_A$ in *I* = 0, i.e. :

$$\frac{1}{(\Delta x)^2}\{[2 + (\Delta x)^2\eta^2(x_1)]\theta_1 - \theta_2\} = T_\infty \eta^2(x_1) + \frac{1}{(\Delta x)^2}T_A \qquad \text{(II.82)}$$

Similarly, for *i* = *N,* the last equation takes into account the condition imposed at the end B of the bar $\theta_{N+1} = T_B$, so that this $N^{ème}$ equation is written as :

$$\frac{1}{(\Delta x)^2}\{-\theta_{N-1} + [2 + (\Delta x)^2\eta^2(x_N)]\theta_N\} = T_\infty \eta^2(x_N) + \frac{1}{(\Delta x)^2}T_B \quad \text{(II.83)}$$

The problem described by the previous linear system can be written in matrix form, with $0_{(\Delta\%)}$ denoting the column vector of unknowns:

$$\theta_{\Delta x} = \begin{vmatrix} \theta_1 \\ \theta_2 \\ \vdots \\ \theta_i \\ \vdots \\ \theta_{N-1} \\ \theta_N \end{vmatrix}$$

By introducing the matrix : $A_{\Delta x}$

$$A_{\Delta x} = \frac{1}{(\Delta x)^2} \begin{vmatrix} 2+(\Delta x)^2\eta^2(x_1) & -1 & 0 & 0 & 0 \\ -1 & 2+(\Delta x)^2\eta^2(x_2) & -1 & 0 & 0 \\ \vdots & \vdots & \vdots & \vdots & \vdots \\ 0 & 0 & -1 & 2+(\Delta x)^2\eta^2(x_{N-1}) & -1 \\ 0 & 0 & 0 & -1 & 2+(\Delta x)^2\eta^2(x_N) \end{vmatrix}$$

And the second member vector $b_{\Delta x}$:

$$b_{\Delta x} = \begin{vmatrix} T_\infty\eta^2(x_1) + \frac{1}{(\Delta x)^2}T_A \\ T_\infty\eta^2(x_2) \\ \vdots \\ T_\infty\eta^2(x_{N-1}) \\ T_\infty\eta^2(x_N) + \frac{1}{(\Delta x)^2}T_B \end{vmatrix}$$

The linear system to be solved is written in matrix form:

$$A_{\Delta x}\theta_{\Delta x} = b_{\Delta x} \qquad \text{(II.84)}$$

The matrix A_{Lx}above contains non-zero terms only on its main diagonal and the two adjacent diagonals. For this reason, it is said to be tridiagonal.

In this case, the general form of the finite difference equation is

$$a_i\theta_{i-1} + b_i\theta_i + c_i\theta_{i+1} = d_i \qquad \text{(II.85)}$$

where the coefficients :

$$a_i = -\frac{1}{(\Delta x)^2}$$

$$b_i = \frac{2}{(\Delta x)^2} + \eta^2(x_i)$$

$$c_i = -\frac{1}{(\Delta x)^2}$$

$$d_i = T_\infty \eta^2(x_i)$$

One method of solving it, known as the tridiagonal matrix algorithm, or AMTD algorithm, involves finding a recurrence relation of the form :

$$\theta_{i-1} = e_i\theta_i + f_i \qquad \text{(II.86)}$$

Transferring this relationship into equation (II.85), we get :

$$(a_i e_i + b_i)\theta_i + c_i\theta_{i+1} = d_i + a_i f_i \qquad \text{(II.87)}$$

from which we get :

$$\theta_i = \frac{-c_i\theta_{i+1} + d_i - a_i f_i}{a_i e_i + b_i} \qquad \text{(II.88)}$$

but the recurrence (II.86) will be written at ranks *i, i+1* :

$$\theta_i = e_{i+1}\theta_{i+1} + f_{i+1} \qquad \text{(II.89)}$$

By identifying relations (II.88) and (II.89), we obtain the following two recurrences, for coefficients *e* and *f* :

$$\begin{cases} e_{i+1} = \dfrac{-c_i}{a_i e_i + b_i} \\ f_{i+1} = \dfrac{d_i - a_i f_i}{a_i e_i + b_i} \end{cases} \qquad \text{(II.90)}$$

This means that if we know the coefficients *e* and *fi* at rank *i*, we can calculate those at the next rank i+1 using the equations above. It is therefore sufficient to know the initial values of *ei* and *fi* to deduce all the other values. These initial values are provided by the first equation in the system to be solved (*i* = 1). This equation is written as :

$$a_1\theta_0 + b_1\theta_1 + c_1\theta_2 = d_1 \qquad \text{(II.91)}$$

where the value *θo* is known and is equal to TA. We therefore have :

$$\theta_1 = \frac{-c_1}{b_1}\theta_2 + \frac{d_1 - a_1 T_A}{b_1} \qquad \text{(II.92)}$$

Equation (II.92) is none other than equation (II.86) for *i*= 2, i.e. :

$$\theta_1 = e_2\theta_2 + f_2 \qquad \text{(II.93)}$$

By identifying equations (II.92) and (II.93), we obtain the initial values *e* 2 and *f2* :

$$\begin{cases} e_2 = \frac{-c_1}{b_1} \\ f_2 = \frac{d_1 - a_1 T_A}{b_1} \end{cases}$$

Knowing the initial values e_2 and $f2$ from equations (II.90), it is possible to calculate all the values of e_i and fi. Once we know all the values of e_i and fi, we can obtain the required solutions θ using equation (II.86). To do this, we start from the temperature T_B imposed at the end B of the bar, which corresponds to the nreud of index N +1. It will therefore be necessary to calculate the values of ei and fi up to rank i = N+1.

It should be noted that equation (II.86) for the linear system corresponding to index N, the general linear form of which is given by equation (II.85), has a special form such that :

$$a_N\theta_{N-1} + b_N\theta_N + 0 = d_N, \quad (c_N = 0) \qquad \text{(II.94)}$$

The recurrence (II.90) therefore imposes :

$$\begin{cases} e_{N+1} = 0 \\ f_{N+1} = \frac{d_N - a_N f_N}{a_N e_N + b_N} \end{cases}$$

This means that the temperature condition imposed at the end B of the bar determines the temperature at the last internal node i = N in the following way:

$$\theta_N = e_{N+1}\theta_{N+1} + f_{N+1} = f_{N+1} \qquad \text{(II.95)}$$

The temperatures at the other internal nodes are calculated successively from BN using the recurrence (II.86).

In order to ensure the validity of a numerical model, it is always desirable to be able to compare the results calculated using the model with a known exact solution for a simple case. So let's return to the mathematical definition (Eq. II.78) of the physical problem for which we have developed a numerical model:

$$\frac{d^2T}{dx^2} - (T - T_\infty)\eta^2(x) = 0$$

It is the $\eta^2(x)$ term that makes it impossible to solve this differential equation. But if this term is independent of x, then there is an exact solution to the equation :

$$\frac{d^2T}{dx^2} - \eta^2(T - T_\infty) = 0$$

It is therefore a second-order differential equation with constant coefficients whose general solution is of the form :

$$T(x) = T_\infty + A_1 \exp(-\eta x) + A_2 \exp(\eta x) \qquad \text{(II.96)}$$

or :

$$T(x) = T_\infty + B_1 \cosh(\eta\, x) + B_2 \sinh(\eta\, x) \qquad \text{(II.97)}$$

The conditions at both ends allow us to determine, in each case, the corresponding solution. For this purpose, equation (II.96) is well suited to problems involving very long bars. Equation (II.97) is suitable for limited geometries.

II.5.1.3.4 Application of the bar: vanes, finned surfaces

The results obtained previously for the bar problem are applicable to fins of different cross-sectional shapes. Obviously, the assumptions of one-dimensional steady-state conduction, no generated source and constant λ and h properties are maintained. In addition, preference will be given to the type of fin most often used, namely constant cross-section fins (Fig. II.22), which are easier to construct and analyse. These fins are used in particular in air conditioners, refrigerators, vehicle and central heating radiators, and for cooling microprocessors (Fig. II.23). They increase the heat exchange between a solid body and the surrounding environment.

In general, for any fin of constant cross-section (Fig. II.22), equation (II.78) is written :

$$\frac{d^2T}{dx^2} - \left\{\frac{hP}{\lambda A_c}\right\}(T - T_\infty) = 0 \qquad \text{(II.98)}$$

By defining an excess temperature such that :

$$\Theta(x) = T(x) - T_\infty \qquad \text{(II.99)}$$

A simple general equation is obtained for the temperature distribution in the fin:

$$\frac{d^2\Theta}{dx^2} - \eta^2\Theta = 0 \qquad \text{(II.100)}$$

a linear, homogeneous equation with constant coefficients whose general solution is :

$$\Theta(x) = C_1 \exp(\eta x) + C_2 \exp(-\eta x) \qquad \text{(II.101)}$$

The expression of the parameter η being :

$$\eta = \sqrt{\frac{h\,P}{\lambda\, A_c}} \qquad \text{(II.102)}$$

The boundary conditions allow us to find the constants c_1 and c_2. In the presence of convection with the surrounding medium, we have :

- At $x = 0$ (base of the temperature fin) T_b): $\Theta_b = T_b - T_\infty$.

At $x =$ L (extreme cross-sectional area of the fin): $h\Theta_L = -\lambda \left.\frac{d\Theta}{dx}\right|_L$.

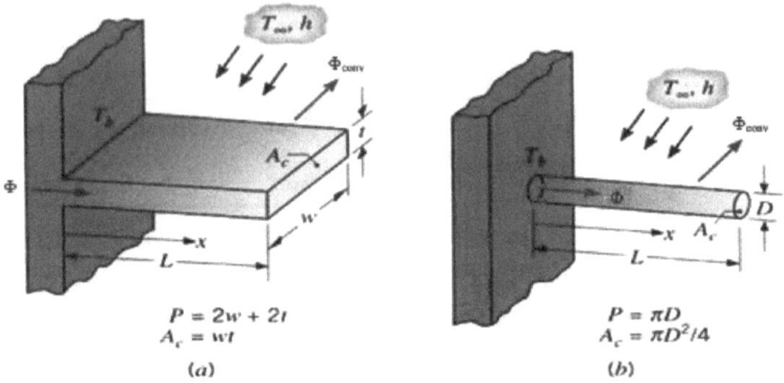

FIGURE II.22: Fin with constant cross-section.

In this case (mixed transfer), the solution is :

$$\frac{\Theta}{\Theta_b} = \frac{\cosh \eta(L-x) + \frac{h}{\eta\,\lambda} \sinh \eta(L-x)}{\cosh(\eta\, L) + \frac{h}{\eta\,\lambda} \sinh(\eta\, L)} \qquad \text{(II. 103)}$$

For an infinitely long fin ($0_L = 0$):

$$\frac{\Theta}{\Theta_b} = \exp(-\eta x) \qquad \text{(II. 104)}$$

■ For the fin whose flux at the tip is zero (insulation), $-\lambda \left.\frac{d\Theta}{dx}\right|_L = 0$:

$$\frac{\Theta}{\Theta_b} = \frac{\cosh \eta(L-x)}{\cosh(\eta\, L)} \qquad \text{(II. 105)}$$

■ For a fin with an imposed tip temperature ($\Theta_L = Cte$):

$$\frac{\Theta}{\Theta_b} = \frac{\left(\frac{\Theta_L}{\Theta_b}\right) \sinh(\eta\, x) + \sinh \eta(L-x)}{\sinh(\eta\, L)} \qquad \text{(II. 106)}$$

Even if the graphical interpretation doesn't really generate any applications, hyperbolic functions are very useful. Simply because they are the solution of the integral of a relatively simple function. A table of these hyperbolic functions can be found at the end of this book (Appendix 3).

To determine the quality of the fin, its actual performance is compared with that of an ideal fin with a uniform temperature equal to the temperature of its base. Such a fin should be made of a material with infinite thermal conductivity. The exchanged flux would be :

$$\Phi_{max} = h\,P\,L\,(T_b - T_\infty) \qquad \text{(II.107)}$$

The efficiency of the fin is then defined by the ratio :

$$\varepsilon = \frac{\Phi_{réel}}{\Phi_{max}} \qquad \text{(II.108)}$$

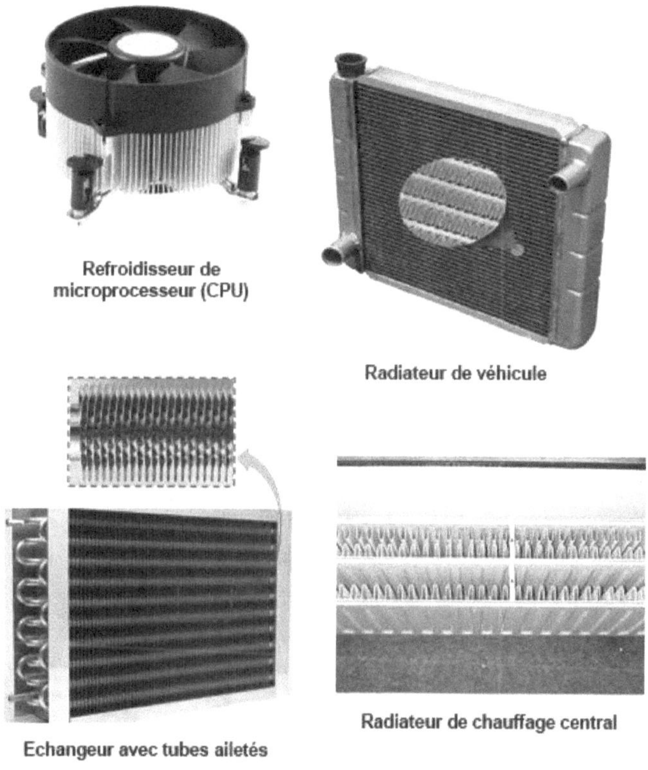

Microprocessor (CPU) cooler Vehicle radiator
Heat exchanger with finned tubes Central heating radiator
FIGURE II.23: Fin applications.

The concept of fin efficiency can be extended to a finned surface (Fig. 24). The exchange surface area of a body, of initial surface area *Si*, is increased by adding a certain number of identical fins of total exchange surface area *Sa*. The efficiency *E* of the finned exchange surface area is therefore defined by the ratio of the flow actually exchanged by the total surface area *St* of the body with its fins (*St* = *Si* + *Sa*) to the maximum flow that could be exchanged by *St* if the whole body were at a uniform temperature equal to the temperature at the base of the fin. In other words :

$$E = \frac{\Phi_{réel}}{\Phi_{max}} = \frac{hS_i\,(T_b - T_\infty) + \varepsilon hS_a\,(T_b - T_\infty)}{hS_t\,(T_b - T_\infty)} = \frac{1 + \varepsilon\left(\frac{S_a}{S_i}\right)}{1 + \left(\frac{S_a}{S_i}\right)} \qquad \text{(II.109)}$$

In summary, fins are added to optimise heat transfer, distribution and dissipation, resulting in increased efficiency, better temperature control and improved safety in a variety of heating applications. Thanks to its additional surfaces, energy savings can be achieved while ensuring efficient heat transfer.

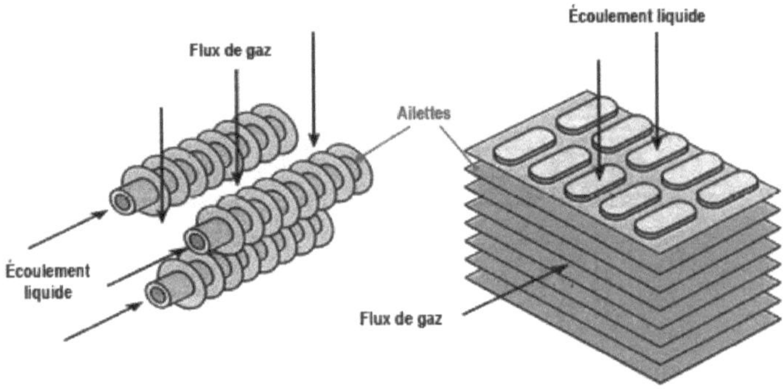

FIGURE II.24: Finned surfaces.

II.5.2 Variable speed

In what follows, we will restrict ourselves to solving thermokinetic problems with two variables: a spatial variable *x* and time *t*. This corresponds to many practical cases where the boundary conditions are independent of *y* and *z*. The heat equation can then be reduced to :

$$\frac{\partial^2 T}{\partial x^2} = \frac{1}{\alpha}\frac{\partial T}{\partial t} \qquad \text{(II.110)}$$

II.5.2.1 Laplace transformation

A first technique for easily solving equation (II.110) above is to apply the Laplace transformation.

II.5.2.1.1 Principle of the method

The Laplace transform $\overline{T}(p)$ of a function $T(t)$ is defined, under certain conditions, by the integral :

$$\overline{T}(x,p) = \int_0^\infty \exp(-pt)\,T(x,t)\,dt \qquad \text{(II.111)}$$

where p is a real or complex number. $T(p)$ is called the image or

transform of the original function *T(t)*.
Applying the Laplace transform (II.111) to the one-dimensional heat equation (II.110) gives :

$$\int_0^{\infty} \exp(-pt)\left(\frac{\partial^2 T}{\partial x^2}\right)dt - \frac{1}{\alpha}\int_0^{\infty} \exp(-pt)\left(\frac{\partial T}{\partial t}\right)dt = 0 \qquad \text{(II.112)}$$

Using the property of differentiation under the sum sign for the first integral and integrating the second by parts, we obtain the equation :

$$\frac{d^2}{dx^2}\int_0^{\infty} \exp(-pt)T(x,t)dt - \frac{1}{\alpha}\left\{\left[T(x,t)\frac{1}{\exp(pt)}\right]_{t=0}^{t=\infty} + p\int_0^{\infty} \exp(-pt)T(x,t)dt\right\} = 0$$

i.e., by the very definition of the Laplace transform :

$$\frac{d^2\overline{T}}{dx^2} - \frac{1}{\alpha}\{0 - T(x,0)\} - \frac{p}{\alpha}\overline{T} = 0$$

By noting To the initial temperature distribution *T*(*x*, 0), the transformed equation is written definitively :

$$\frac{d^2\overline{T}}{dx^2} - \frac{p}{\alpha}\overline{T} = \frac{T_0}{\alpha} \qquad \text{(II.113)}$$

We can now see the whole point of the Laplace transform, which is to reduce a partial differential equation whose solution seemed impossible to obtain to a simple second-order differential equation, which is easier to solve.
Given $T^* = T - T_0$, , the transformed equation reduces to a differential equation without a second member: whose general solution is :

$$\frac{d^2\overline{T}^*}{dx^2} - \frac{p}{\alpha}\overline{T}^* = 0 \qquad \text{(II.114)}$$

whose general solution is :

$$\overline{T}^*(x,p) = A\exp(-kx) + B\exp(kx) \qquad \text{(II.115)}$$

with : $k^2 = \frac{p}{\alpha}$

From the transformed solution $\overline{T}^*(x,p)$, we arrive at the original solution $T^*(x,t) = T(x,t) - T_0$ using the inverse transformation. In practice, it is simpler to use the results given by the transform tables. Table (II.3) gives some Laplace transforms which will be used in the example below to illustrate the effectiveness of this solving technique.

TABLE II.3: Some Laplace transforms.

$T(t)$	$\overline{T}(p) = \int_0^{\infty} \exp(-pt)T(t)dt$
1	$\frac{1}{p}$
$aT_1(t) + bT_2(t)$	$a\overline{T}_1(p) + b\overline{T}_2(p)$
$\frac{\partial T(t)}{\partial x}$	$p\overline{T}(p) - T(0)$
$\frac{\partial^n T}{\partial x^n}$	$\frac{d^n \overline{T}}{dx^n}$
$\operatorname{erfc}\left(\frac{x}{2\sqrt{\alpha t}}\right)$	$\frac{\exp(-kx)}{p}$
$2\sqrt{\frac{\alpha t}{\pi}} \exp\left(\frac{-x^2}{4\alpha t}\right) - x \operatorname{erfc}\left(\frac{x}{2\sqrt{\alpha t}}\right)$	$\frac{\exp(-kx)}{kp}$

II.5.2.1.2 Application of the Laplace transform: Semi-infinite wall subjected to a temperature jump

The medium defined by half-space is called a semi-infinite wall. A simple example is a plane floor whose surface can be subjected to various conditions.

Consider such a semi-infinite wall, initially at a uniform temperature T_0 throughout its mass (Fig. II.25). It is assumed that its surface is suddenly heated to a constant temperature T_1.

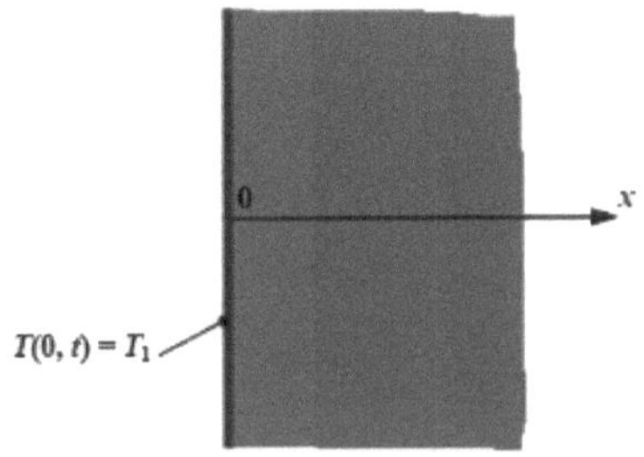

FIGURE II.25: Semi-infinite wall.

The system to be solved to calculate the temperature distribution *T*(*x*, *t*) in the wall is written :

$$\begin{cases} \dfrac{\partial^2 T}{\partial x^2} - \dfrac{1}{\alpha}\dfrac{\partial T}{\partial t} = 0 \\ T(0,t) = T_1 \\ T(x,0) = T_0 \end{cases}$$

By making the change of variable $T^*(x,t) = T(x,t) - T_0$ indicated previously, the system to be solved is then written :

$$\begin{cases} \dfrac{\partial^2 T^*}{\partial x^2} - \dfrac{1}{\alpha}\dfrac{\partial T^*}{\partial t} = 0 \\ T^*(0,t) = T_1 - T_0, \text{température imposée à la surface du mur.} \\ T^*(x,0) = 0, \text{distribution de température initiale à travers le mur.} \end{cases}$$

temperature imposed on the surface of the wall.
initial temperature distribution across the wall.

By applying the Laplace transformation to the previous system, we obtain the system :

$$\begin{cases} \dfrac{d^2\overline{T}^*}{dx^2} - \dfrac{p}{\alpha}\overline{T}^* = 0 \\ \overline{T}^*(0,p) = \dfrac{T_1 - T_0}{p} \end{cases}$$

whose general solution is given by expression (II.115).

Since the temperature *T*(*x*, *t*) and its transform $\overline{T}(x,p)$ cannot tend towards infinity, the integration constant *B* is necessarily zero. The temperature condition at *x* = 0 imposes $A = \frac{(T_1 - T_0)}{p}$, , hence the expression for the transformed solution :

$$\overline{T}^*(x,p) = (T_1 - T_0)\frac{\exp(-kx)}{p}$$

From table (II.3), the original solution *T'*(*x*, *t*) is given by the inverse transform of the function : $\overline{T}(x,p)$

$$T^*(x,t) = (T_1 - T_0)\,\text{erfc}\left(\frac{x}{2\sqrt{\alpha t}}\right)$$

Taking into account the initial change of variable :

$$\frac{T(x,t) - T_0}{T_1 - T_0} = \text{erfc}(u), \text{avec } u = \frac{x}{2\sqrt{\alpha t}} \qquad \text{(II.116)}$$

The function erfc(*u*) is the complementary error function defined by :

erfc(u) = 1 - erf(u)

The error function erf(*u*) is itself defined by the integral :

$$\operatorname{erf}(u) = \frac{2}{\sqrt{\pi}} \int_0^u \exp(-t^2)\, dt$$

This classic function is used not only in thermal engineering, but also in statistics, hence its name. Its use is crucial for analysing data distributions and understanding variations within a set. It allows researchers and statisticians to interpret results more accurately and meaningfully. The erf(u) and erfc(u) functions are given at the end of this book, in the tables in Appendix 2.

II.5.2.2 Semi-infinite wall in sinusoidal regime

The problems of heat conduction in solids whose temperature undergoes periodic variations come into play in many cases, such as :

- Space heating, taking into account heat gain due to sunshine.
- Calculation of temperature fields in internal combustion engines, determined by the periodic nature of the heat flow.
- Automatic temperature control.

We will study the simple case of a semi-infinite wall subjected to a sinusoidal variation in the temperature of its face at x = 0. After a sufficiently long time, the temperature of the medium will evolve around a constant mean value at each point of abscissa x. We are interested in the fluctuation 6(x, t) of the temperature around this mean value, in what is known as an established sinusoidal regime.

The system of equations governing this problem is as follows

$$\begin{cases} \dfrac{\partial^2 \theta}{\partial x^2} - \dfrac{1}{\alpha}\dfrac{\partial \theta}{\partial t} = 0 \\ \theta(0,t) = \theta_0 \cos \omega t \\ \theta(\infty, t) \to 0 \end{cases} \qquad \text{(II. 117)}$$

The temperature of the face x = 0 of the wall oscillates sinusoidally with a pulsation ω. We will look for a solution that is also sinusoidal, with the same pulsation ω.

We are looking for a particular solution, with separate variables x and t, satisfying the heat equation of the system (II.117) of the form:

$$\theta(x,t) = X(x).Y(t) \qquad \text{(II. 118)}$$

Transferring (II.118) into (II.117), we obtain :

$$\frac{X''(x)}{X(x)} = \frac{1}{\alpha}\frac{Y'(t)}{Y(t)} \qquad \text{(II. 119)}$$

The first member of equation (II.119) does not depend on t. The second member does not depend on x. Since the equality of the two members must be verified for all values of x and t, this is only possible if each member is equal to a constant, i.e. :

$$\frac{\alpha X''(x)}{X(x)} = \beta \qquad \text{(II.120a)}$$

$$\frac{Y'(t)}{Y(t)} = \beta \qquad \text{(II.120b)}$$

The choice of constant β will depend on the thermal process being studied. By choosing an imaginary number $\beta = i\omega$, we will arrive at a solution that is suitable for describing phenomena that are periodic as a function of time. Depending on the nature of the physical problem, the real or imaginary part of the complex solution obtained will be retained.

Equation (II.120a) is in the form :

$$X''(x) - \frac{\beta}{\alpha} X(x) = 0$$

which is a second order differential equation whose general solution is of the form :

$$X(x) = A_1 \exp\left(-\sqrt{\frac{\beta}{\alpha}}\, x\right) + A_2 \exp\left(\sqrt{\frac{\beta}{\alpha}}\, x\right) \qquad \text{(II.121)}$$

The solution of equation (II.120b) is :

$$Y(t) = A_3 \exp^{(() \beta t)}$$

so the complex solution we are looking for is written as :

$$\theta_i(x,t) = \left\{A_1 \exp\left(-\sqrt{\frac{\beta}{\alpha}}\, x\right) + A_2 \exp\left(\sqrt{\frac{\beta}{\alpha}}\, x\right)\right\} A_3 \exp(\beta t) \qquad \text{(II.122)}$$

A_1, A_2 and A_3 are three integration constants. For the solution to remain finite when x tends towards infinity, A_2 must necessarily = 0. On the $x = 0$ wall we must have :

$$\theta_i(0,t) = (A_1 A_3) \exp(i\omega t) = \theta_0 (\cos \omega t + \sin \omega t) \qquad \text{(II.123)}$$

which imposes the value 60 for the integration constant ($A1A_3$). Note that :

$$\sqrt{\beta} = \sqrt{i\omega} = \sqrt{\frac{\omega}{2}}\,(1+i)$$

The complex solution (II.122) can be written as :

$$\theta_i(x,t) = \theta_0 \exp\left(-\sqrt{\frac{\omega}{2\alpha}}\,(1+i)x\right) \exp(i\omega t) \qquad \text{(II.124)}$$

Finally, the temperature fluctuation we are looking for is the real part of the above complex solution, i.e. :

$$\theta(x,t) = \theta_0 \exp\left(-\sqrt{\frac{\omega}{2\alpha}}\,x\right)\cos\left(\omega t - \sqrt{\frac{\omega}{2\alpha}}\,x\right) \qquad (II.125)$$

II.5.2.3 Quenching a metal ball

This process consists of immersing a ball, initially at temperature *Ti*, in a bath maintained at constant temperature To, (*To * Ti).*

It is assumed that the temperature of the ball is uniform, which will almost certainly be the case if the ball is small and/or if the thermal conductivity of the material making up the ball is high.

In this problem, there is no internal production of heat (medium with no internal sources). The energy conservation equation is written :

$$m\,C\,\frac{dT}{dt} = \Phi \qquad (II.126)$$

The heat flow exchanged across the boundary here is a convective flow governed by Newton's law. The heat flux density is such that :

$$\vec{\varphi} = h\,(T_p^{bille} - T_0)\vec{n}$$

The temperature of the bilie being uniform, $T_p^{bille} = T$ over the entire surface *S* of the ball. Therefore :

$$\Phi = \iint -(\vec{\varphi}\cdot\vec{n})\,dS = -h\,S\,(T(t) - T_0) \qquad (II.127)$$

Equation (II.126) is then written :

$$m\,C\,\frac{dT}{dt} = -h\,S\,(T(t) - T_0)$$

i.e. :

$$\frac{d(T(t) - T_0)}{dt} = -\frac{h\,S}{m\,C}\,(T(t) - T_0) \qquad (II.128)$$

assuming: $\tau = \dfrac{m\,C}{h\,S}$ a thermal time constant. *i L kJ*

which implies :

$$\frac{d(T - T_0)}{(T - T_0)} = -\frac{1}{\tau}\,dt$$

after integration :

$$T - T_0 = A\exp\left(-\frac{t}{\tau}\right)$$

Given the initial condition, *T*(*t*=*0*) = *Ti*, we deduce the constant *A* :

$$A = T_i - T_0$$

so :

$$\frac{T - T_0}{T_i - T_0} = \exp\left(-\frac{t}{\tau}\right) \qquad (II.129)$$

and for $t = \tau$:

$$\frac{T(\tau) - T_0}{T_i - T_0} = \exp(-1) = 0{,}3679$$

This means that at time τ, the difference between the temperature of the bimetallic layer and the temperature of the medium is about 37% of the initial difference.

When $t \to \infty$, , the asymptotic regime is reached and we have :

$$\frac{T - T_0}{T_i - T_0} \to 0 \Leftrightarrow T \to T_0$$

The temperature of the ball tends towards the temperature of the medium in which it is immersed: at equilibrium, the system (ball + medium) is asymptotically isothermal.

The characteristic parameter of the problem that allows us to verify the validity of the uniform-temperature medium approximation is a dimensionless number called the Biot number (*Bi*), defined as the ratio between two thermal resistances:

$$Bi = \frac{R_{int}}{R_{ext}} \qquad \text{(II. 130)}$$

Rint is the internal resistance, related to the ball :

$$R_{int} = \frac{\Delta T_{int}}{\Phi_{int}}$$

Rext is the external resistance, related to the fluid:

$$R_{ext} = \frac{\Delta T_{ext}}{\Phi_{ext}}$$

By replacing the resistances in the Biot equation and indicating *Tp* as the external wall temperature of the ball, we obtain :

$$Bi = \frac{\Phi_{ext}}{\Phi_{int}} \frac{\Delta T_{int}}{\Delta T_{ext}} = \frac{\Phi_{ext}}{\Phi_{int}} \frac{T - T_p}{T_p - T_0} \qquad \text{(II. 131)}$$

- $Bi \ll 1$ si $T \to T_p$ Tp : uniform temperature in the solid (very good good conductors).
- $Bi \gg 1$ si $T_p \to T_0$ Tp : the fluid imposes its temperature on the surface of the solid.

The thermal resistances for conduction and convection, respectively, are :

$$R_{int} = \frac{L_c}{\lambda S} \; et \; R_{ext} = \frac{1}{h S}$$

where *Lc* = *V/S* is a characteristic length of the solid system under

consideration. For a metal ball (a solid sphere of radius *r*), *Lc* = r/3.
Biot's number is expressed as :

$$Bi = \frac{h\,L_c}{\lambda} \qquad \text{(II.132)}$$

Let's take the expression for the time constant *ý* of the system under study:

$$\tau = \frac{m\,C}{h\,S} = \frac{\rho\,V\,C}{h\,S} = \frac{\rho\,L_c\,S\,C}{h\,S} = \frac{\rho\,C\,\lambda\,L_c^2}{h\,\lambda\,L_c} = \frac{L_c^2}{\alpha}\,\frac{1}{Bi}$$

where $\alpha = \frac{\lambda}{\rho\,C}$ is the thermal diffusivity of the balance.
In short :

$$\frac{t}{\tau} = Bi\,\frac{t}{L_c^2/\alpha} \qquad \text{(II.133)}$$

L_c^2/α corresponds to a thermal diffusion time. Physically, it expresses the penetration of heat inside the solid.
An adimensional time is introduced for transient problems, known as the Fourier number (*Fo*).

$$Fo = \frac{\alpha\,t}{L_c^2} \qquad \text{(II.134)}$$

Equation (II.129) becomes :

$$\frac{T - T_0}{T_i - T_0} = \exp[-(Bi\,Fo)] \qquad \text{(II.135)}$$

II.5.2.4 Numerical solution of the heat equation by finite difference methods

In the fairly frequent case where the geometry of the system and the boundary conditions are complex, then numerical methods are well suited to solving heat conduction problems.
Solving the steady-state heat equation (section II.5.1.1) involves transforming the heat equation and the boundary conditions into a system of finite difference equations. The variables are discretised by dividing up the area under study using a mesh with dimensions adapted to each problem.
The difference equations can be established according to what is known as an explicit or implicit finite difference scheme. In other words, the calculation of the temperature at time *t+Δt*, denoted $T^{+A}t$, can be carried out in different ways depending on whether the Laplacian is expressed from the known values T^t of the temperature at time *t* (explicit method) or from the unknown values $T^{+A}t$ of the temperature at time *t+Δt* (implicit

method).

III.5.2.4.1 Explicit scheme

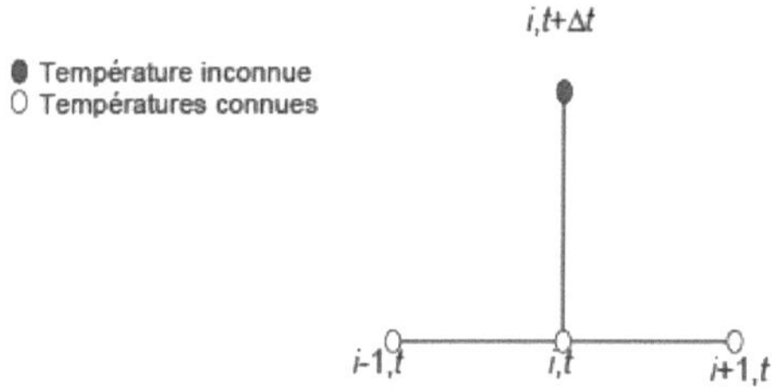

Unknown temperature

Known temperatures

For a one-dimensional heat flow, in the case of a medium with no internal source and constant thermal characteristics, the heat equation (known as the Fourier equation) is :

$$\frac{\partial^2 T}{\partial x^2} = \frac{1}{\alpha}\frac{\partial T}{\partial t}$$

this equation is expressed in terms of finite differences for each of the internal nodes :

$$\frac{T_{i+1}^{t} + T_{i-1}^{t} + 2T_i^t}{\Delta x^2} = \frac{1}{\alpha}\left[\frac{T_i^{t+\Delta t} - T_i^t}{\Delta t}\right] \qquad \text{(II.136)}$$

where the only unknown is given by :

$$T_i^{t+\Delta t} = Fo(T_{i+1}^{t} + T_{i-1}^{t}) + (1 - 2Fo)T_i^t \qquad \text{(II.137)}$$

Ou $Fo = \frac{\alpha\,\Delta t}{\Delta x^2}$.

The advantage of this technique is that each equation contains only one unknown, but in the explicit scheme the choice of space and time increments plays an important role in the stability of the solution. These parameters cannot be chosen arbitrarily and there are limit values for which the solution is stable. Indeed, given that the temperatures and *Fo* are positive quantities, the coefficient of T_i^t cannot be negative, as this would require the value of the temperature at time *t+Δt* to be all the lower as its value at time *t* is greater. This situation gives rise to instability in the calculation, resulting in oscillations in temperature values that increase in magnitude. Consequently, in order for the diagram to be stable at all times, having chosen Δ *t*, the following stability condition for the system of finite difference equations must be met:

$$1 - 2\alpha \frac{\Delta t}{\Delta x^2} \geq 0$$

$$\Rightarrow \alpha \frac{\Delta t}{\Delta x^2} \leq \frac{1}{2}$$

leading to the maximum permissible value of the time step :

$$\Delta t_{adm} = \frac{\Delta x^2}{2\alpha} \qquad \text{(II.138)}$$

In the fairly frequent case of a mixed condition (Fig. II.26), the heat balance, for a unit area element normal to the *x-axis* and of thickness *Δχ/2*, is written :

$$h\,(T_\infty - T_1^t) - \lambda \left(\frac{T_1^t - T_2^t}{\Delta x} \right) = \rho\, C\, \frac{\Delta x}{2} \left(\frac{T_1^{t+\Delta t} - T_1^t}{\Delta t} \right) \qquad \text{(II.139)}$$

Solving the equation in an explicit form with respect to the unknown function $T_1^{t+\Delta t}$, we obtain:

$$T_1^{t+\Delta t} = 2\,BiFo\,T_\infty + (1 - 2\,BiFo - 2\,Fo)T_1^t + 2\,Fo\,T_2^t \qquad \text{(II.140)}$$

In order to set the stability conditions, the coefficients assigned to the temperatures T_1^t and T_2^t must be positive. Consequently :

$$\frac{\alpha\,\Delta t}{\Delta x^2}\left(1 + \frac{h\,\Delta x}{\lambda}\right) \leq \frac{1}{2} \Rightarrow Fo(1 + Bi) \leq \frac{1}{2}$$

$$Bi = \frac{h\,\Delta x}{\lambda}.$$

or

This gives the maximum permissible value for the time step :

$$\Delta t_{adm} = \frac{0{,}5}{\left(\frac{\alpha}{\Delta x^2}\right) + \left(\frac{h}{\rho c\,\Delta x}\right)} \qquad \text{(II. 141)}$$

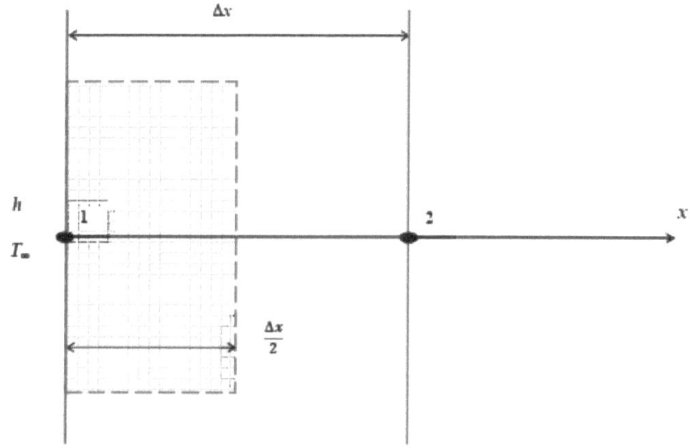

FIGURE II.26: Mixed condition at the boundary node.

III.5.2.4.2 Implicit scheme

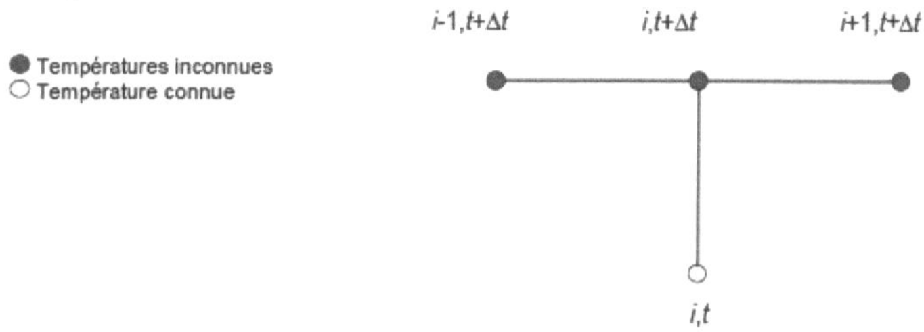

Unknown temperatures

Known temperature

To improve the accuracy of the resolution, we must choose Δx small enough. However, in the explicit scheme, the maximum permissible value of Δt is proportional to $\Delta x^{(2)}$ (stability condition). In these circumstances, it may turn out that, in order to complete the solving process, a considerable number of time steps must be taken, making solving practically impossible. The heat equation is then solved using implicit finite difference equations of the form :

$$\frac{T_{i+1}^{t+\Delta t} + T_{i-1}^{t+\Delta t} - 2T_i^{t+\Delta t}}{\Delta x^2} = \frac{1}{\alpha}\left(\frac{T_i^{t+\Delta t} - T_i^t}{\Delta t}\right) \qquad \text{(II. 142)}$$

Let

$$\left(\frac{1}{Fo}+2\right)T_i^{t+\Delta t}-T_{i+1}^{t+\Delta t}-T_{i-1}^{t+\Delta t}=\frac{1}{Fo}T_i^t \qquad \text{(II.143)}$$

The implicit method uses "backward" differences and offers the advantage of always being stable for any value of *Fo*, so the space and time increments can be chosen without any particular constraints. The explicit method, on the other hand, uses "forward" differences and imposes the stability criterion. In addition, the implicit scheme provides more accurate results than those obtained by the explicit method. However, the procedure for solving the system of equations obtained using this scheme is complicated insofar as, for each increment of time, the equation contains the Liois unknowns: $T_{i-1}^{t+\Delta t}, T_i^{t+\Delta t}, T_{i+1}^{t+\Delta t}$.

In general, it is also possible to express the heat equation for an internal node by the following finite difference equation :

$$\frac{1}{\Delta x^2}\{\delta(T_{i+1}^t+T_{i-1}^t-2T_i^t)+(1-\delta)(T_{i+1}^{t+\Delta t}+T_{i-1}^{t+\Delta t}-2T_i^{t+\Delta t})\}$$

$$=\frac{1}{\alpha\,\Delta t}\left(T_i^{t+\Delta t}-T_i^t\right) \qquad \text{(II.144)}$$

where δ is a coefficient such that: . $0 \le \delta \le 1$.

It turns out that, for the particular values below :

- $\delta=0$ $\rightarrow$ implicit method,
- $\delta=0{,}5$ $\rightarrow$ Crank-Nicholson method,
- $\delta=1$ $\rightarrow$ explicit method.

We have only dealt with variable conduction problems in the simplistic case of a one-dimensional medium. As soon as the geometry of the systems studied becomes more complicated, whether in two or three dimensions, any analytical solution becomes excessively cumbersome and precarious. The only effective tool is a whole range of specialised numerical calculation codes, the description and theory of which are beyond the scope of this book.

APPLICATIONS

A1. A cylindrical electrical resistor (D = 0.4 cm, L = 1.5 cm) on a printed circuit board dissipates a power Φ = 0.6 W. Assuming the heat is transferred uniformly across all surfaces. Determine:

a) The amount of heat dissipated by this resistor over a 24 hour period.

b) The surface heat flux density of the resistor, in W/cm^2.

c) The fraction of heat dissipated by the top and bottom surfaces. Conclude.

Solution

a) Heat dissipated in 24 hours :

$$Q = \Phi * 24 = 14{,}4\ \text{Wh} = 51{,}84\ \text{kJ}$$

b) Flux density :

$$S = 2\frac{\pi D^2}{4} + \pi DL = 0{,}251 + 1{,}885 = 2{,}136\ \text{cm}^2$$

$$\Rightarrow \varphi = \frac{\Phi}{S} = \frac{0{,}6}{2{,}136} = 0{,}2809\ \text{W/cm}^2$$

c) Q_{h-b} is the heat transferred through the top and bottom surfaces. S_{h-b} :

$$\frac{Q_{h-b}}{Q} = \frac{S_{h-b}}{S} = \frac{0{,}251}{2{,}136} = 0{,}118 \text{ soit } 11{,}8\%$$

The fraction of heat dissipated through the top and bottom surfaces is small. Most of the heat transfer takes place through the cylindrical surface.

A2. A refrigerator in the shape of a rectangular parallelepiped 120 cm high, 60 cm wide and 50 cm deep. The walls are made of plastic sheets, thickness e = 3 mm and thermal conductivity λ = 0.12 W/(m.K). The average temperature on the outside is 20°C.

a) Calculate the power required by the refrigeration unit to maintain the average temperature of the inside of the refrigerator at 5°C.

b) What would happen to this power if the walls of the refrigerator were made of two 3 mm thick plastic sheets separated by a 4 cm thick layer of glass wool? The thermal conductivity of glass wool is ALV = 0.04 W/(m.K).

Assume that heat loss is uniform across all six sides of the refrigerator.

Solution

a) Power required :

$$S_T = 2(S_1 + S_2 + S_3) = 2(0{,}72 + 0{,}60 + 0{,}30) = 3{,}24\ \mathrm{m^2}$$

$$R = \frac{e}{\lambda\, S_T} = \frac{0{,}003}{0{,}12 * 3{,}24} = 0{,}0077\ \mathrm{K/W}$$

Hence :

$$\Phi = \frac{T_1 - T_2}{R} = \frac{15}{0{,}0077} = 1944\ \mathrm{W}$$

b) New power :

$$R_{LV} = \frac{0{,}04}{0{,}04 * 3{,}24} = 0{,}3086\ \mathrm{K/W}$$

$$R_T = 2R + R_{LV} = 0{,}324\ \mathrm{K/W}$$

Hence

$$\Phi = \frac{T_1 - T_2}{R_T} = \frac{15}{0{,}324} = 46{,}3\ \mathrm{W}$$

Heat loss is reduced by 97.6%.

A3: Consider a composite wall made of horizontal planks (**P1**) measuring 50 mm x 100 mm and spaced 450 mm apart, to which 25 mm thick planks (**P2)** are fixed to form the outside of the wall and 10 mm thick fibre cement boards (**FC**) form the inside. This wall is completed with glass wool (**LV**) placed between the **P1** boards (see figure with details).

a) Draw up the electrical model.
b) Determine the equivalent thermal resistance.
c) Determine the heat flow through this wall.

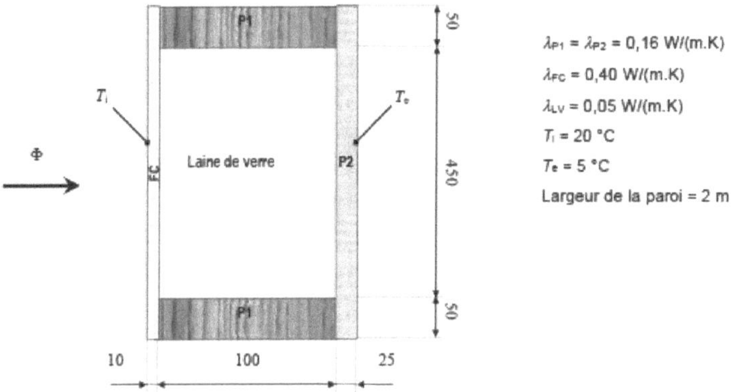

Wall width = 2 m

Solution

a) Electrical model :

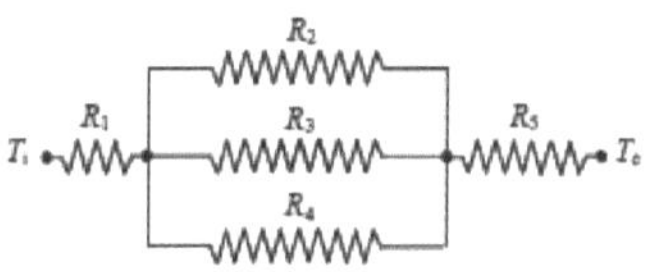

b) Equivalent thermal resistance :

$$R_{eq} = R_1 + \frac{1}{\left(\frac{1}{R_2} + \frac{1}{R_3} + \frac{1}{R_4}\right)} + R_5$$

With,

$$R_1 = \frac{e_{FC}}{\lambda_{FC}\, S_{FC}} = \frac{0{,}010}{0{,}40 * (0{,}550 * 2)} = 0{,}023\ \mathrm{K/W}$$

$$R_2 = R_4 = \frac{e_{P1}}{\lambda_{P1}\, S_{P1}} = \frac{0{,}100}{0{,}16 * (0{,}050 * 2)} = 6{,}250\ \mathrm{K/W}$$

$$R_3 = \frac{e_{LV}}{\lambda_{LV}\, S_{LV}} = \frac{0{,}100}{0{,}05 * (0{,}450 * 2)} = 2{,}222\ \mathrm{K/W}$$

$$R_5 = \frac{e_{P2}}{\lambda_{P2}\, S_{P2}} = \frac{0{,}025}{0{,}16 * (0{,}550 * 2)} = 0{,}142\ \mathrm{K/W}$$

Numerical application :

$$R_{eq} = 0{,}023 + \frac{1}{\left(\frac{1}{6{,}250} + \frac{1}{2{,}222} + \frac{1}{6{,}250}\right)} + 0{,}142 = 1{,}464\ \mathrm{K/W}$$

c) Heat flow :

$$\Phi = \frac{T_i - T_e}{R_{eq}} = \frac{20 - 5}{1{,}464} = 10{,}25 \text{ W}$$

A4. Steam at 145°C flows through a steel pipe (*Лac* = *45* W/(m.K)), 48 mm internal diameter and 56 mm external diameter, covered with asbestos insulation (*Àam* = 0.174 W/(m.K)), 75 mm external diameter. The thermal resistance of the inner wall due to vapour is 0.172 K/W and the ambient temperature is 21°C. The thermal resistance at the outer wall (air) is negligible.

a) Calculate the heat flux lost per unit length.

b) Calculate the

.

Solution

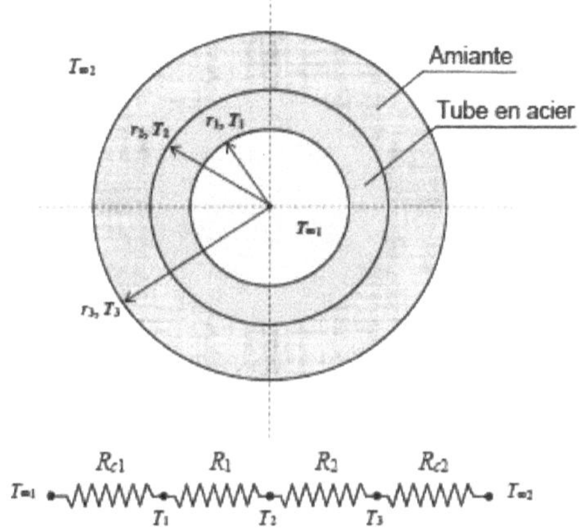

a) Heat flux lost per metre of length :

$$\Phi = \frac{T_{\infty 1} - T_{\infty 2}}{R_T} = \frac{T_{\infty 1} - T_{\infty 2}}{R_{c1} + R_1 + R_2 + R_{c2}}$$

With,

$$R_1 = \frac{\ln \frac{r_2}{r_1}}{2\,\pi\,\lambda_{ac} L} = \frac{\ln \frac{28}{24}}{2 * \pi * 45 * 1} = 0{,}0005\ \mathrm{K/W}$$

$$R_2 = \frac{\ln \frac{r_3}{r_2}}{2\,\pi\,\lambda_{am} L} = \frac{\ln \frac{37{,}5}{28}}{2 * \pi * 0{,}174 * 1} = 0{,}2672\ \mathrm{K/W}$$

$$\Rightarrow \Phi = \frac{145 - 21}{0{,}172 + 0{,}0005 + 0{,}2672 + 0} = 282\ \mathrm{W/m}$$

b) Temperatures T1, T2 and T3 :

$$T_1 = T_{\infty 1} - (R_{c1}\Phi) = 145 - (0{,}172 * 282) = 96{,}50\ °\mathrm{C}$$

$$T_2 = T_1 - (R_1\,\Phi) = 96{,}5 - (0{,}0005 * 282) = 96{,}36\ °\mathrm{C}$$

$$T_3 = T_{\infty 2} = 21\ °\mathrm{C}\ \text{(Résistance convective extérieure négligeable)}$$

(External convective resistance negligible)

A5. A hollow cylinder, with thermal conductivity λ = 70 W/(m.K), whose cylindrical surfaces are isothermal, is heated on the inside. It is not possible to measure the inside temperature T_1 at r_1 and the temperature T_2 at r_2. The temperature T at r and the flux density φ at r_2 are known. We ask you to calculate the temperatures T_1 and T_2 on the inside and outside of the cylinder.

Assume steady-state heat conduction with no internal heat source .

Solution

$r_1 = 0{,}04$ m

$r_2 = 0{,}08$ m

$\varphi(r_2) = 20\times10^4$ W/m²

$T = 400$ °C

$r = 0{,}06$ m

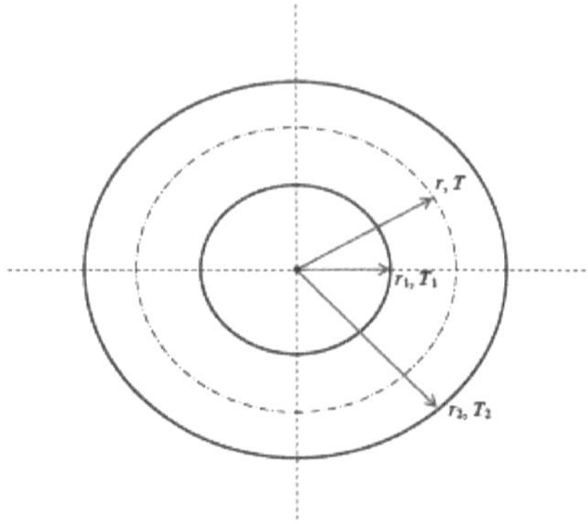

- Calculate temperatures T1 and T2 :

$$T(r) = T_1 + \frac{T_1 - T_2}{\ln\left(\frac{r_1}{r_2}\right)} \ln\left(\frac{r}{r_1}\right) \Rightarrow \varphi(r) = -\lambda\,\frac{dT(r)}{dr} = -\lambda\,\frac{T_1 - T_2}{\ln\left(\frac{r_1}{r_2}\right)}\frac{1}{r}$$

We have the system :

$$\begin{cases} T(r = 0{,}06) = T_1 + \dfrac{T_1 - T_2}{\ln\left(\frac{r_1}{r_2}\right)} \ln\left(\dfrac{r}{r_1}\right) = 400\ °\text{C} \rightarrow (1) \\ \varphi(r_2) = -\lambda\,\dfrac{T_1 - T_2}{\ln\left(\frac{r_1}{r_2}\right)}\dfrac{1}{r_2} = 20 \times 10^4\ \text{W} \rightarrow (2) \end{cases}$$

$(2) \Rightarrow T_1 - T_2 = 158{,}5$ °C

$(1) \Rightarrow T_1 = 492{,}7$ °C

Hence

$T_2 = 334{,}2$ °C

A6. The wall of a furnace is made of refractory bricks with a thickness *e* = 250 mm. The temperature of its inner wall is T_1 = 1350°C and the temperature of its outer wall is T_2 = 50°C. The thermal conductivity of the bricks varies linearly with the temperature Ts according to the law: Λ= *ЛО* (1 + *PT)*. a)Deduce the expression for the heat flux transmitted through the wall of the furnace.

b)Calculate this flux per m²of surface and for: *ЛО* = 0.838 W/(m.K) and *0* = 0.0007 K⁻¹.

c) Find the temperature distribution law *T*(*x*) in the wall. Validate at the boundaries of the wall.

Solution

a) Expression of heat flow :

$$\Phi = -\lambda(T)\, S\, \frac{dT}{dx} = -\lambda_0(1+\beta T)\, S\, \frac{dT}{dx}$$

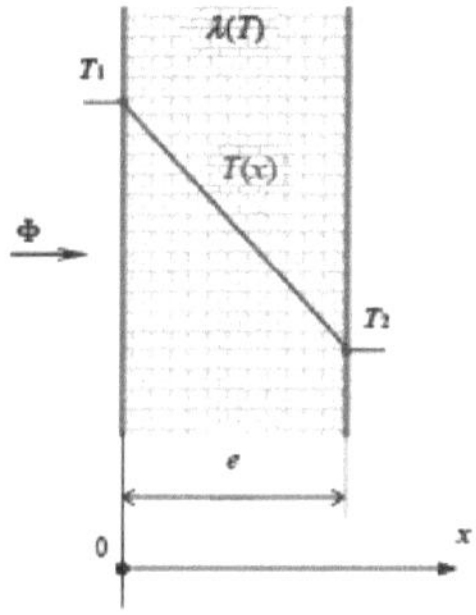

This equation is integrated, after separating the variables *T* and *x*, over the thickness of the oven wall:

$$\frac{\Phi}{S}\int_0^e dx = -\lambda_0 \int_{T_1}^{T_2} (1+\beta T)\, dT$$

$$\Rightarrow \Phi = \frac{\lambda_0\, S}{e}\,(T_1 - T_2)\left[1 + \frac{\beta}{2}(T_1 + T_2)\right]$$

b) And for a surface area of 1 m2 :

$$\Phi = \frac{0{,}838}{0{,}250} * 1300\left(1 + \frac{0{,}0007}{2} * 1400\right) = 6492{,}824\ \mathrm{W/m^2}$$

c) To obtain the temperature field *T*(*x*) inside the oven wall, we repeat the same integration as in (a) but between 0 and *x* along the oven wall, i.e. between T1 and *T*(*x*), for temperatures :

$$\frac{\Phi}{S}\int_0^x dx = -\lambda_0 \int_{T_1}^{T(x)} (1+\beta T)\, dT$$

$$\Rightarrow \frac{\Phi\, x}{S\, \lambda_0} = T_1 - T(x) + \frac{\beta}{2}\, T_1^2 - \frac{\beta}{2}\, T^2(x)$$

Second-degree equation with solution :

$$T(x) = -\frac{1}{\beta} + \sqrt{\frac{1}{\beta^2} - \frac{2\Phi}{\beta\, S\, \lambda_0}x + \frac{2T_1}{\beta} + T_1^2}$$

$$= -\frac{1}{\beta} + \sqrt{\left(\frac{1}{\beta} + T_1\right)^2 - \frac{2\Phi}{\beta\, S\, \lambda_0}x}$$

This expression validates both boundary conditions:

$$En\ x = 0 \longrightarrow T(x=0) = 1350\ ^{\circ}\mathrm{C}$$

$$En\ x = e \longrightarrow T(x=e) = 50\ ^{\circ}\mathrm{C}$$

<u>**A7**</u>. Assume a plate, thickness *e* = 10 cm, insulated on side 1 and a

temperature *T0* imposed on side 2. Assuming that the plate generates heat by volume at constant flux *pv* = 106 W/m3 , for a constant thermal conductivity:

a)Give the expression for the temperature of the plate $T(x)$.

b) Calculate the temperature of side 1.

Given: λ = 40 W/(m.K).

Solution

a) Expression of the plate temperature $T(x)$:

In this case, the one-dimensional heat equation is Poisson's.

$$\frac{d^2T}{dx^2} + \frac{p_v}{\lambda} = 0 \text{ avec } 0 < x < e$$

$$\Rightarrow \frac{dT}{dx} = -\frac{p_v}{\lambda}x + C_1$$

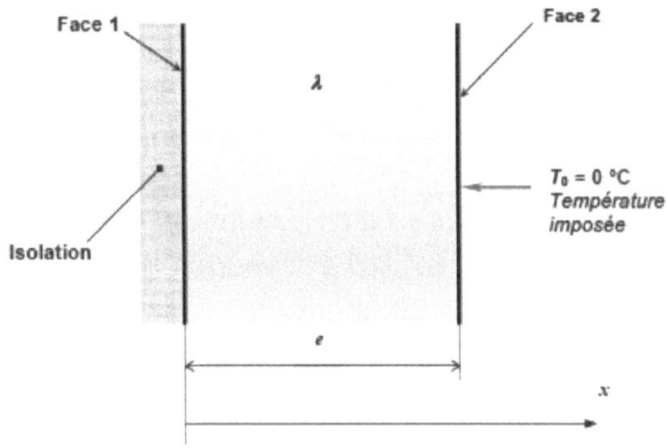

$$\Rightarrow T(x) = -\frac{1}{2}\frac{p_v}{\lambda}x^2 + C_1 x + C_2$$

C. L. :

$$\begin{cases} pour\ x = 0: \dfrac{dT(x)}{dx} = 0 \Rightarrow C_1 = 0 \\ pour\ x = e: T(e) = T_0 = 0\,°\text{C} \Rightarrow T(e) = -\dfrac{1}{2}\dfrac{p_v}{\lambda}e^2 + C_2 \Rightarrow C_2 = \dfrac{1}{2}\dfrac{p_v}{\lambda}e^2 \end{cases}$$

To sum up:

$$T(x) = -\frac{1}{2}\frac{p_v}{\lambda}x^2 + \frac{1}{2}\frac{p_v}{\lambda}e^2 = \frac{1}{2}\frac{p_v}{\lambda}[e^2 - x^2]$$

b) Temperature of side 1 (x = 0) :

$$T(x=0) = \frac{1}{2}\frac{p_v}{\lambda}e^2 = \frac{1}{2}\frac{10^6}{40}\,0{,}1^2 = 125\,°\text{C}$$

A8. A plate of thickness *e* = 6 mm, made of a material with a thermal conductivity coefficient *λ* = 20 W/(m.K), has a uniformly distributed internal heat source *pv* = 5x10^{7}W/m^3. The temperatures at the boundary surfaces of the plate are T_1 = 120°C and T_2 = 127.2°C respectively.

a) Find the expression for the temperature *T*(*x*).

b) Calculate the maximum temperature in the plate.

c) Calculate the heat flux density at the boundary surfaces. The system is steady-state with constant *λ* and *pv*.

Solution

a) Function *T*(*x*) :

$$T(x) = -\frac{1}{2}\frac{p_v}{\lambda}x^2 + C_1 x + C_2$$

En *x* = 0 → $C_2 = T_1$

En *x* = e → *T* = T_2, d'où :

$$T(x=e) = T_2 = -\frac{1}{2}\frac{p_v}{\lambda}e^2 + C_1 x + T_1 \Rightarrow C_1 = \frac{T_2 - T_1}{e} + \frac{p_v\, e}{2\lambda}$$

By replacing the constants c_1 and c_2 found in *T*(*x*) :

$$T(x) = -\frac{1}{2}\frac{p_v}{\lambda}x^2 + \left[\frac{T_2 - T_1}{e} + \frac{p_v\, e}{2\lambda}\right]x + T_1$$

b) Maximum temperature :

$$\frac{dT}{dx} = -\frac{p_v}{\lambda}x + C_1 = 0 \Rightarrow x = \frac{C_1\,\lambda}{p_v}$$

hence :

$$T_{max} = -\frac{p_v}{2\lambda}\left(\frac{C_1\,\lambda}{p_v}\right)^2 + C_1\left(\frac{C_1\,\lambda}{p_v}\right) + T_1$$

$$\Rightarrow T_{max} = -\frac{p_v}{2\lambda}\left[\frac{\lambda}{p_v}\left\{\frac{T_2 - T_1}{e} + \frac{p_v\, e}{2\lambda}\right\}\right]^2 + \left[\frac{T_2 - T_1}{e} + \frac{p_v\, e}{2\lambda}\right]^2\frac{\lambda}{p_v} + T_1$$

$$\Rightarrow T_{max} = -15{,}138 + 30{,}276 + 120 = 135{,}1\ °\mathrm{C}$$

c) Flux density :

$$\varphi = -\lambda\frac{dT}{dx} = -\lambda\left(-\frac{p_v}{\lambda}x + C_1\right) = p_v\, x - C_1\,\lambda$$

$$pour\ x = 0 \longrightarrow \varphi_1 = -C_1\,\lambda = \frac{\lambda\,(T_1 - T_2)}{e} - \frac{p_v\, e}{2} = -174\ \mathrm{kW}$$

$$pour\ x = e \longrightarrow \varphi_2 = p_v\, e - C_1\,\lambda = p_v\, e + \frac{\lambda\,(T_1 - T_2)}{e} - \frac{p_v\, e}{2}$$

$$\varphi_2 = 126\ \mathrm{kW}$$

A9. An aluminium fin (λ = 200 W/(m.K)) with a rectangular cross-section 3 mm thick and 7.5 cm long protrudes from a wall (see figure below). Its base is maintained at 300°C and the ambient temperature is 50°C with h = 10 W/(m$^{(2)}$·K). Calculate :

a) The heat flux dissipated by the fin to the surrounding environment per unit width of material.

b) The efficiency of this fin.

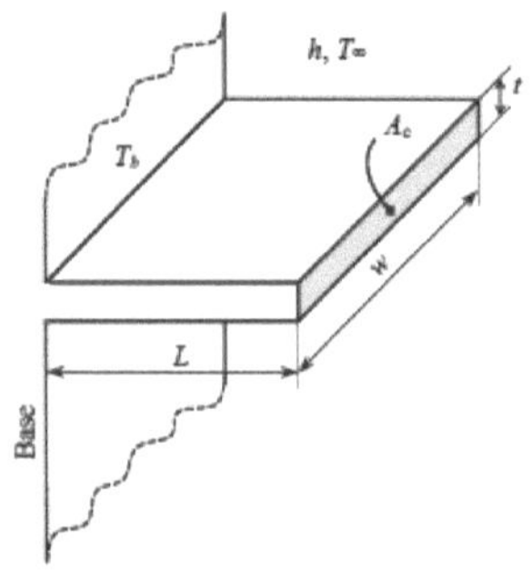

Solution

a) The convection flux dissipated on the lateral surface of the blade and the area of its extreme right section is, in the present steady state case, equal to the flux transferred by conduction
in the right section of the fin at its base (x = 0).
In other words :

$$\Phi = -\lambda S \left.\frac{dT}{dx}\right|_{x=0} = \left\{\int_0^L h P\,(T(x) - T_\infty)\,dx\right\} + h A_c(T(L) - T_\infty)$$

The temperature distribution is given by :

$$\frac{T(x) - T_\infty}{T_b - T_\infty} = \frac{\cosh \eta(L - x) + \frac{h}{\eta \lambda} \sinh \eta(L - x)}{\cosh(\eta L) + \frac{h}{\eta \lambda} \sinh(\eta L)}$$

$$\Rightarrow \Phi = -\lambda S \left.\frac{dT}{dx}\right|_{x=0}$$

$$= \lambda \eta A_c(T_b - T_\infty) \frac{\sinh(\eta L) + \frac{h}{\eta \lambda} \cosh(\eta L)}{\cosh(\eta L) + \frac{h}{\eta \lambda} \sinh(\eta L)}$$

$$\text{ou bien: } \Phi = \lambda\, \eta\, A_c (T_b - T_\infty)\, \frac{\tanh(\eta\, L) + \frac{h}{\eta\, \lambda}}{1 + \frac{h}{\eta\, \lambda} \tanh(\eta\, L)}$$

$$\text{avec: } \eta = \sqrt{\frac{h\, P}{\lambda\, A_c}} = \sqrt{\frac{h\, 2(w+t)}{\lambda\, (w\, t)}} = \sqrt{\frac{10 * 2(1 + 0{,}003)}{200 * (1 * 0{,}003)}}$$

$$= 5{,}782 \text{ m}^{-1}$$

Numerical application :

$$\Phi = 200 * 5{,}782 * 0{,}003 * 250\, \frac{\tanh(0{,}434) + 0{,}009}{1 + 0{,}009 \tanh(0{,}434)}$$

$$= 361 \text{ W/m}$$

b) Fin efficiency :

$$\varepsilon = \frac{\Phi}{\Phi_{max}} = \frac{\lambda\, \eta\, A_c (T_b - T_\infty) \left(\frac{\tanh(\eta\, L) + \frac{h}{\eta\, \lambda}}{1 + \frac{h}{\eta\, \lambda} \tanh(\eta\, L)} \right)}{h\, P\, L (T_b - T_\infty)}$$

$$\Rightarrow \varepsilon = \frac{\tanh(\eta\, L) + \frac{h}{\eta\, \lambda}}{(\eta\, L) + \frac{h\, L}{\lambda} \tanh(\eta\, L)}$$

Numerical application :

$$\varepsilon = \frac{\tanh(0{,}434) + 0{,}009}{(0{,}434) + 0{,}00375 * \tanh(0{,}434)} = 0{,}959 \; (\varepsilon = 95{,}9\%)$$

A10. a) At what minimum depth should a water pipe be buried so that a sudden drop in temperature to -15°C does not cause the pipe to freeze after 15 days? the dry soil has a uniform initial temperature of 5°C, its thermal diffusivity is given by :

$\alpha = 2{,}8 \times 10^{-7}$ m²/s.

b) At what depth L must we penetrate into the ground for the amplitude of the temperature fluctuation, at this depth, to be no more than 1/100th of the amplitude at the surface over a period of 24 hours? It is assumed that the daily variations in temperature at the surface of the ground are sinusoidal.

Solution

a) The solution to this problem is given by the semi-infinite wall equation:

$$\frac{T(x,t) - T_0}{T_1 - T_0} = \text{erfc}\left(\frac{x}{2\sqrt{at}}\right)$$

$$\Rightarrow T(x,t) = T_0 + (T_1 - T_0)\,\text{erfc}\left(\frac{x}{2\sqrt{at}}\right)$$

With an initial ground temperature *T0* of 5°C, the pipe will freeze at depth *x* and time *t* when the condition :

T(x, t) = 0

Condition satisfied when :

$$\text{erfc}\left(\frac{x}{2\sqrt{at}}\right) = 1 - \text{erf}\left(\frac{x}{2\sqrt{at}}\right) = \frac{-T_0}{T_1 - T_0} = \frac{-5}{-20} = 0{,}25$$

$$\Rightarrow \text{erf}\left(\frac{x}{2\sqrt{at}}\right) = 1 - 0{,}25 = 0{,}75$$

From the table of the erfc (*u*) function in Appendix 2, we derive the solution :

$$\text{u} = \frac{x}{2\sqrt{at}} = 0{,}8$$

That is, after 15 days ($1.296\chi10^6$seconds):

$$x = 0{,}8 * 2\sqrt{2{,}8\times10^{-7} * 1{,}296\times10^{6}} = 0{,}964 \text{ m} \cong 1 \text{ m}$$

Buried at a depth of 1 m, the water pipe will take 15 days to cool down from +5 °C to 0 °C, when the temperature at the surface of the ground drops unexpectedly from +5 °C to -15 °C.

b) The amplitude of temperature fluctuations at depth *x* = *L*, has the value :

$$\theta(x = L) = \theta_0 \exp\left(-\sqrt{\frac{\omega}{2\alpha}}\,L\right)\cos\left[\omega t - \sqrt{\frac{\omega}{2\alpha}}\,L\right]$$

and that at the surface (*x* = 0) :

$$\theta(x = 0) = \theta_0\,\cos\,(\omega t)$$

The desired depth *L* will therefore be given by the condition :

$$\frac{\theta(x = L)}{\theta(x = 0)} = \exp\left(-\sqrt{\frac{\omega}{2\alpha}}\,L\right) = \frac{1}{100} \Rightarrow \sqrt{\frac{\omega}{2\alpha}}\,L = \ln 100$$

d'où :

$$L = \frac{4{,}605}{\sqrt{\frac{\omega}{2\alpha}}}$$

As the period of the fluctuation is 24 hours, we can deduce the pulsation :

$$\omega = \frac{2\pi}{\mathrm{T}} = \frac{2\pi}{24 * 3600} = 7{,}272\times10^{-5}\ \mathrm{s}^{-1}$$

Finally, the depth *L* we are looking for has the value :

$$L = \frac{4{,}605}{\sqrt{\frac{7{,}272\times10^{-5}}{2 * 2{,}8\times10^{-7}}}} = 0{,}404\ \mathrm{m}$$

BIBLIOGRAPHICAL REFERENCES

1. SACADURA J.F.: "Initiation aux Transferts Thermiques". 445 p., Lavoisier Technique et Documentation, Paris, 1980.
2. HOLMAN J.P.: "Heat Transfer". 4th Edition, 530 p., McGraw-Hill Book Company, U.S.A., 1976.
3. CARSLAW H.S., JAEGER J.C.: "Conduction of Heat in Solids". 2nd Edition, 520 p., Oxford University Press, U.S.A., 1986.
4. MZAD H.: "Applied Heat Transfer: Spray Cooling". 116 p., European University Publishing, Germany, 2024.
5. TALER J., DUDA P. Solving Direct and Inverse Heat Conduction Problems". 915 p., Springer, Germany, 2006.
6. BATTAGLIA J.L., KUSIAK A., PRADÈRE C. Introduction aux Transferts Thermiques". Cours et exercices corrigés. 3rd Edition, 320 p., Dunod, France, 2020.
7. HAHN D.W., ÖZI§IK M.N.: "Heat Conduction". 3rd Edition, 752 p., John Wiley & Sons, Inc, U.S.A., 2012.
8. BENICHOU P., BENICHOU R., BOY N., POUGET J.P., DEMENGEL G.: "Séries de Fourier - Transformation de Laplace". 224 p., Ellipses, France, 1995.
9. DEGIOVANNI A.: "Transmission de l'Énergie Thermique - Conduction". Techniques de l'Ingénieur, BE8200 v1, France, 1999.
10. JEANNOT Y., MOYNE C., DEGIOVANNI A.: "Transfert de Chaleur 1 : Conduction et Convection". 322 p., ISTE Éditions, France, 2022.
11. TRIBOIX A., BOUVENOT J.B.: "Transferts Thermiques : Méthodes Numérique". 2nd Edition, 382 p., Eyrolles, France, 2022.
12. LE HIR J.: "Thermodynamique: Transferts Thermiques". Tome 3, 184 p. Éditions Cépaduès, France, 2024.

Mass heat capacity of various elements.
Standard conditions at TAPN, i.e. 25°C and 101.3 kPa.

Name	Formula	Heat by mass, J/(g.°c)	Name	Formula	Heat of mass, J/(g.°c)
Aluminium	Al	0,897	Krypton	Kr	0,248
Silver	Ag	0,235	Magnesium	Mg	1,023
Argon	Ar	0,520	Manganese	Mn	0,479
Nitrogen	N_2	1,040	Mercury	Hg	0,140
Beryllium	Be	1,825	Neon	Ne	1,030
Boron	B	1,026	Nickel	Ni	0,444
Carbon (graphite)	C	0,709	Octosulphur	S_8	0,710
Dibromium	Br_2	0,474	Gold	Au	0,129
Chlorine	Cl_2	0,479	Platinum	Pt	0,133
Difluor	F_2	0,824	Lead	Pb	0,129
Hydrogen	H_2	14,304	Radon	Rn	0,094
Diiodine	I_2	0,214	Scandium	Sc	0,568
Dioxygen	O_2	0,918	Selenium	Se	0,321
Chromium	Cr	0,449	Silicon	Si	0,705
Cobalt	Co	0,421	Tetraphosphorus	P_4	0,769
Copper	Cu	0,385	Titanium	Ti	0,523
Tin	Sn	0,228	Tungsten	W	0,132
Iron	Fe	0,449	Uranium	U	0,116
Gallium	Ga	0,371	Vanadium	V	0,489
Germanium	Ge	0,320	Xenon	Xe	0,158
Helium	He	5,193	Zinc	Zn	0,388

Appendix 1 (continued)

Heat capacity by mass of some liquid substances.
Standard conditions at TAPN, i.e. 25°C and 101.3 kPa.

Name	Formula	Heat by mass, J/(g.°c)
Antifreeze	-	2,20
Water	H_2O	4,19
Ethanol	C_2H_6O	2,50
Ether	-	2,20
Ethylene glycol	$C_2H_6O_2$	2,40
Hexane	C_6H_{14}	2,27
Oil	-	1,29 - 2,0
Mercury	Hg	0,139
Octane	C_8H_{18}	0,140

Error function table.

$$\text{erf}\ (u) = \frac{2}{\sqrt{\pi}} \int_0^u \exp(-t^2)\ dt$$

	Chiffre des centièmes de *u*									
u	**0**	**1**	**2**	**3**	**4**	**5**	**6**	**7**	**8**	**9**
0,0	0,00000	0,01128	0,02256	0,03384	0,04511	0,05637	0,06762	0,07886	0,09008	0,10128
0,1	0,11246	0,12362	0,13476	0,14587	0,15695	0,16800	0,17901	0,18999	0,20094	0,21184
0,2	0,22270	0,23352	0,24430	0,25502	0,26570	0,27633	0,28690	0,29742	0,30788	0,31828
0,3	0,32863	0,33891	0,34913	0,35928	0,36936	0,37938	0,38933	0,39921	0,40901	0,41874
0,4	0,42839	0,43797	0,44747	0,45689	0,46623	0,47548	0,48466	0,49375	0,50275	0,51167
0,5	0,52050	0,52924	0,53790	0,54646	0,55494	0,56332	0,57162	0,57982	0,58792	0,59594
0,6	0,60386	0,61168	0,61941	0,62705	0,634459	0,64203	0,64938	0,65663	0,66378	0,67084
0,7	0,67780	0,68467	0,69143	0,69810	0,70468	0,71116	0,71754	0,72382	0,73001	0,73610
0,8	0,74210	0,74800	0,75381	0,75952	0,76514	0,77067	0,77610	0,78144	0,78669	0,79184
0,9	0,79691	0,80188	0,80677	0,81156	0,81627	0,82089	0,82542	0,82987	0,83423	0,83851
1,0	0,84270	0,84681	0,85084	0,85478	0,85865	0,86244	0,86614	0,86977	0,87333	0,87680
1,1	0,88021	0,88353	0,88679	0,88997	0,89308	0,89612	0,89910	0,90200	0,90484	0,90761
1,2	0,91031	0,91296	0,91553	0,91805	0,92051	0,92290	0,92524	0,92751	0,92973	0,93190
1,3	0,93401	0,93606	0,93807	0,94002	0,94191	0,94376	0,94556	0,94731	0,94902	0,95067
1,4	0,95229	0,95385	0,95538	0,95686	0,95830	0,95970	0,96105	0,96237	0,96365	0,96490
1,5	0,96611	0,96728	0,96841	0,96952	0,97059	0,97162	0,97263	0,97360	0,97455	0,97546
1,6	0,97635	0,97721	0,97804	0,97884	0,97962	0,98038	0,98110	0,98181	0,98249	0,98315
1,7	0,98379	0,98441	0,98500	0,98558	0,98613	0,98667	0,98719	0,98769	0,98817	0,98864
1,8	0,98909	0,98952	0,98994	0,99035	0,99074	0,99111	0,99147	0,99182	0,99216	0,99248
1,9	0,99279	0,99309	0,99338	0,99366	0,99392	0,99418	0,99443	0,99466	0,99489	0,99511
2,0	0,99532	0,99552	0,99572	0,99591	0,99609	0,99626	0,99642	0,99658	0,99673	0,99688
2,1	0,99702	0,99715	0,99728	0,99741	0,99753	0,99764	0,99775	0,99785	0,99795	0,99805
2,2	0,99814	0,99822	0,99831	0,99839	0,99846	0,99854	0,99861	0,99867	0,99874	0,99880
2,3	0,99886	0,99891	0,99897	0,99902	0,99906	0,99911	0,99915	0,99920	0,99924	0,99928
2,4	0,99931	0,99935	0,99938	0,99941	0,99944	0,99947	0,99950	0,99952	0,99955	0,99957
2,5	0,99959	0,99961	0,99963	0,99965	0,99967	0,99969	0,99971	0,99972	0,99974	0,99975
2,6	0,99976	0,99978	0,99979	0,99980	0,99981	0,99982	0,99983	0,99984	0,99985	0,99986
2,7	0,99987	0,99987	0,99988	0,99989	0,99989	0,99990	0,99991	0,99991	0,99992	0,99992
2,8	0,99992	0,99993	0,99993	0,99994	0,99994	0,99994	0,99995	0,99995	0,99995	0,99996
2,9	0,99996	0,99996	0,99996	0,99997	0,99997	0,99997	0,99997	0,99997	0,99997	0,99998
3,0	0,99998	0,99998	0,99998	0,99998	0,99998	0,99998	0,99998	0,99999	0,99999	0,99999
3,1	0,99999	0,99999	0,99999	0,99999	0,99999	0,99999	0,99999	0,99999	0,99999	0,99999
3,2	0,99999	0,99999	0,99999	1,00000	1,00000	1,00000	1,00000	1,00000	1,00000	1,00000

Complementary error function table.

$$\text{erfc}\ (u) = 1 - \text{erf}\ (u) = \frac{2}{\sqrt{\pi}} \int_u^{\infty} \exp(-t^2)\, dt$$

	Chiffre des centièmes de *u*									
u	0	1	2	3	4	5	6	7	8	9
0,0	1,00000	0,98872	0,97744	0,96616	0,95489	0,94363	0,93238	0,92114	0,90992	0,89872
0,1	0,88754	0,87638	0,86524	0,85413	0,84305	0,83200	0,82099	0,81001	0,79906	0,78816
0,2	0,77730	0,76648	0,75570	0,74498	0,73430	0,72367	0,71310	0,70258	0,69212	0,68172
0,3	0,67137	0,66109	0,65087	0,64072	0,63064	0,62062	0,61067	0,60079	0,59099	0,58126
0,4	0,57161	0,56203	0,55253	0,54311	0,53377	0,52452	0,51534	0,50625	0,49725	0,48833
0,5	0,47950	0,47076	0,46210	0,45354	0,44506	0,43668	0,42838	0,42018	0,41208	0,40406
0,6	0,39614	0,38832	0,38059	0,37295	0,36541	0,35797	0,35062	0,34337	0,33622	0,32916
0,7	0,32220	0,31533	0,30857	0,30190	0,29532	0,28884	0,28246	0,27618	0,26999	0,26390
0,8	0,25790	0,25200	0,24619	0,24048	0,23486	0,22933	0,22390	0,21856	0,21331	0,20816
0,9	0,20309	0,19812	0,19323	0,18844	0,18373	0,17911	0,17458	0,17013	0,16577	0,16149
1,0	0,15730	0,15319	0,14916	0,14522	0,14135	0,13756	0,13386	0,13023	0,12667	0,12320
1,1	0,11979	0,11647	0,11321	0,11003	0,10692	0,10388	0,10090	0,09800	0,09516	0,09239
1,2	0,08969	0,08704	0,08447	0,08195	0,07949	0,07710	0,07476	0,07249	0,07027	0,06810
1,3	0,06599	0,06394	0,06193	0,05998	0,05809	0,05624	0,05444	0,05269	0,05098	0,04933
1,4	0,04771	0,04615	0,04462	0,04314	0,04170	0,04030	0,03895	0,03763	0,03635	0,03510
1,5	0,03389	0,03272	0,03159	0,03048	0,02941	0,02838	0,02737	0,02640	0,02545	0,02454
1,6	0,02365	0,02279	0,02196	0,02116	0,02038	0,01962	0,01890	0,01819	0,01751	0,01685
1,7	0,01621	0,01559	0,01500	0,01442	0,01387	0,01333	0,01281	0,01231	0,01183	0,01136
1,8	0,01091	0,01048	0,01006	0,00965	0,00926	0,00889	0,00853	0,00818	0,00784	0,00752
1,9	0,00721	0,00691	0,00662	0,00634	0,00608	0,00582	0,00557	0,00534	0,00511	0,00489
2,0	0,00468	0,00448	0,00428	0,00409	0,00391	0,00374	0,00358	0,00342	0,00327	0,00312
2,1	0,00298	0,00285	0,00272	0,00259	0,00247	0,00236	0,00225	0,00215	0,00205	0,00195
2,2	0,00186	0,00178	0,00169	0,00161	0,00154	0,00146	0,00139	0,00133	0,00126	0,00120
2,3	0,00114	0,00109	0,00103	0,00098	0,00094	0,00089	0,00085	0,00080	0,00076	0,00072
2,4	0,00069	0,00065	0,00062	0,00059	0,00056	0,00053	0,00050	0,00048	0,00045	0,00043
2,5	0,00041	0,00039	0,00037	0,00035	0,00033	0,00031	0,00029	0,00028	0,00026	0,00025
2,6	0,00024	0,00022	0,00021	0,00020	0,00019	0,00018	0,00017	0,00016	0,00015	0,00014
2,7	0,00013	0,00013	0,00012	0,00011	0,00011	0,00010	0,00009	0,00009	0,00008	0,00008
2,8	0,00008	0,00007	0,00007	0,00006	0,00006	0,00006	0,00005	0,00005	0,00005	0,00004
2,9	0,00004	0,00004	0,00004	0,00003	0,00003	0,00003	0,00003	0,00003	0,00003	0,00002
3,0	0,00002	0,00002	0,00002	0,00002	0,00002	0,00002	0,00002	0,00001	0,00001	0,00001
3,1	0,00001	0,00001	0,00001	0,00001	0,00001	0,00001	0,00001	0,00001	0,00001	0,00001
3,2	0,00001	0,00001	0,00001	0,00000	0,00000	0,00000	0,00000	0,00000	0,00000	0,00000

Table of hyperbolic functions.

$$\sinh(x) = \frac{1}{2}[\exp(x) - \exp(-x)]$$

$$\cosh(x) = \frac{1}{2}[\exp(x) + \exp(-x)]$$

x	sinh *x*	cosh *x*	tanh *x*	*x*	sinh *x*	cosh *x*	tanh *x*
0,0	0,0000	1,0000	0,0000	2,3	4,9370	5,0372	0,9801
0,1	0,1002	1,0050	0,0997	2,4	5,4662	5,5569	0,9837
0,2	0,2013	1,0201	0,1974	2,5	6,0502	6,1323	0,9866
0,3	0,3045	1,0453	0,2913	2,6	6,6947	6,7690	0,9890
0,4	0,4108	1,0811	0,3800	2,7	7,4063	7,4735	0,9910
0,5	0,5211	1,1276	0,4621	2,8	8,1919	8,2527	0,9926
0,6	0,6367	1,1855	0,5371	2,9	9,0596	9,1146	0,9940
0,7	0,7586	1,2552	0,6044	3,0	10,018	10,068	0,9951
0,8	0,8881	1,3374	0,6640	3,5	16,543	16,573	0,9982
0,9	1,0265	1,4331	0,7163	4,0	27,290	27,308	0,9993
1,0	1,1752	1,5431	0,7616	4,5	45,003	45,014	0,9998
1,1	1,3356	1,6685	0,8005	5,0	74,203	74,210	0,9999
1,2	1,5095	1,8107	0,8337	5,5	122,34	122,35	1,0000
1,3	1,6984	1,9709	0,8617	6,0	201,71	201,72	1,0000
1,4	1,9043	2,1509	0,8854	6,5	332,57	332,57	1,0000
1,5	2,1293	2,3524	0,9052	7,0	548,32	548,32	1,0000
1,6	2,3756	2,5775	0,9217	7,5	904,02	904,02	1,0000
1,7	2,6456	2,8283	0,9354	8,0	1490,5	1490,5	1,0000
1,8	2,9422	3,1075	0,9468	8,5	2457,4	2457,4	1,0000
1,9	3,2682	3,4177	0,9562	9,0	4051,5	4051,5	1,0000
2,0	3,6269	3,7622	0,9640	9,5	6679,9	6679,9	1,0000
2,1	4,0219	4,1443	0,9705	10,0	11013	11013	1,0000
2,2	4,4571	4,5679	0,9757				